HyperWars

*Eleven Essential Strategies for Survival and Profit
in the Era of Online Business*

BRUCE JUDSON

with Kate Kelly

A TOUCHSTONE BOOK
PUBLISHED BY SIMON & SCHUSTER

TOUCHSTONE
1230 Avenue of the Americas
New York, NY 10020

First Touchstone Edition 2000

TOUCHSTONE and colophon are trademarks of Simon & Schuster Inc.

DESIGNED BY ERICH HOBBING

Manufactured in the United States of America

10 9 8 7 6 5 4 3 2 1

The Library of Congress has cataloged the Scribner edition as follows:

Judson, Bruce.
Hyperwars: eleven strategies for survival and profit in the era of
online business/Bruce Judson with Kate Kelly.
p.   cm.
Includes index.
1. Business enterprises—Computer networks. 2. Internet (Computer network)
3. Electronic commerce. 4. Internet marketing.
I. Kelly, Kate. II. Title.
HD30.37.J83   1999
658'.054678—dc21   98-47790 CIP

ISBN 0-684-85564-X
ISBN 0-684-85565-8 (Pbk)

This book is dedicated
to the memory of my father,
MORRIS JUDSON,
who never saw a mountain
he did not encourage me to climb.

Acknowledgments

The authors would like to thank the many people who contributed to this project. They include: our literary agents, Stuart Krichevsky and Judith Riven, who believed in the book from the start; our editor and publisher at Scribner, Gillian Blake and Susan Moldow, who helped us to create the best possible book; our respective spouses, Nancy Judson and George Schweitzer, who bore the brunt of the long hours involved in this project; Michael Katz, who made this collaboration possible; Ross Jaffe of Brentwood Venture Capital, who made the phone call discussed in the Introduction; and Larry Kirshbaum, Bill Sarnoff, Marshall Loeb, and Michael Wolff, who all provided invaluable guidance in the creation of this book.

The authors would like to thank as well the many individuals—who are too numerous to list—working in new media firms, technology companies, and brick-and-mortar businesses who shared their experiences and perspectives with us.

Contents

Why I Wrote
This Book

At Time Warner in the mid-1990s, I became one of the early Internet pioneers, working as a leader of the team that created one of the first large-scale commercial Internet sites. From that time, my work has kept me in the midst of a swirl of Internet development, and my experiences, both positive and negative, have now spanned the full spectrum of online activities: I was the cofounder of Time Warner's Pathfinder, which, for many years, was one of the highest traffic Internet sites in the world; I was involved in the start of Road Runner, the high-speed cable Internet access service; I served on the board of directors of several of the leading Internet commerce and marketing companies; I have been extensively involved with a variety of Internet technology companies; and I wrote the best-selling book *NetMarketing*, which was one of the first books to discuss the business potential of the Internet. I have also launched one of the first paid online newsletters, *Bruce Judson's Grow Your Profits*, which details how businesses can use the Internet to save money, save time, and profitably grow their revenues.

With this wide experience in business and the Internet, I decided to write this book for one simple reason: I was constantly being asked by people from all walks of life—from sole proprietors and independent agents to CEOs of major companies—about how the Internet would affect them. Everywhere I went, I was asked questions like these:

- How will my business or industry be changed?
- Will I still have a business?
- Will I still have a job?
- How will my job change?
- What should I do now?

This book is written to provide practical answers to these questions.

I would also like to note that this is not a book about the Internet. It is book about business, written by a businessperson for businesspeople. As the Internet is rapidly redirecting the currents of all businesses, this book is a nautical chart for navigating the potentially difficult waters ahead.

Observations from the Front Lines

My perspective on how the Internet will affect businesses is shaped by my involvement in its early and ongoing commercialization. At this point, I feel that the Internet has already matured through several different phases, each of which has launched the need for a new business battle plan. The end result is that companies have forever been changed because of the new types of competitive dynamics at play. Now, the stakes are even higher because the Internet has become accessible to more and more people, making an ever-increasing number of businesses and individuals dependent on the Web.

For me, the start of each new phase has been marked by a particularly telling watershed event:

The first was Netscape's unexpected decision in 1994 to release its browser free on the Internet, which confirmed my belief that we were entering a new world. In its earliest days, when I was general manager of new media at Time, Inc., Netscape was working on a browser to sell to the general public. One afternoon I was meeting with Jim Clark, the chairman and cofounder of Netscape, in the Time Inc. boardroom. We were discussing how Netscape planned to build its business and what type of role Time Warner might play in it. Clark paused for a moment and then looked at me and said, "It's confidential, but tomorrow we are going to release the first version of our browser for free to everyone on the Internet."

I looked up in surprise and immediately responded: "But then how will you create a moneymaking business?"

He smiled somewhat enigmatically and after a pause said, "I'm sure we'll find a way."

The immediate implications of Netscape's actions were clear. First, it meant that millions of people would have access to a powerful new browser, thereby ensuring that use of the Internet would explode. Second, Netscape's decision to use "free" to almost instantly build a customer base of millions of people had the potential of forever changing the nature of competition.

The second watershed event occurred during a trip to San Francisco in 1996. Late in the day, a friend, who was a partner at a large venture capital firm, called me and said, "There's something highly confidential we want to show you. Time Warner might want to be involved. The only time we can do it is at nine P.M., but it's worth it. Can you make it?"

That night I was driven down multiple dark roads, only somewhat facetiously wondering if I would soon be blindfolded for extra security. Ultimately, we stopped at a nondescript building and went inside the door of a company named Artemis Research, where I was introduced to Steve Perlman, the CEO of the company. His first words to me were "The real name of our company is WebTV."

After I saw WebTV demonstrated that night, months before it was made public, I knew that the Internet would soon pass through another phase. This type of technological progress guaranteed that one way or the other, consumers would have inexpensive access to the Web. I had long believed Web use was likely to be as ubiquitous as the telephone, now I was certain. (A few weeks later, I also figured out a possible meaning of the company's public name, Artemis: In the 1960s, a popular television show, *The Wild Wild West,* featured a quick-change artist, Artemis Gordon. Artemis Research was similarly not what it appeared to be.)

Perlman's fanatical approach to secrecy was telling as well. Later in this book, I discuss the value of speed. It's clear that Steve felt coming to market first was so important that he didn't want anyone else to know what he and his team were doing in the event potential competitors might then accelerate their work or steal the publicity—valuable for sales momentum—if they had developments of their own.

The third watershed event, and yet another phase in the development of the Internet as a potent business force, occurred when America Online went offline for nineteen hours in August 1996. Businesses across the United States were affected by the loss of e-mail; progress of the "repair" was covered by radio and television almost moment by moment; and it was front-page news.

I subsequently shared a speaking panel on the West Coast with Ted Leonsis, the president of AOL Studios, and Halsey Minor, the CEO and founder of CNET. Stewart Alsop, a well-known technology journalist and *Fortune* columnist, was the moderator. At some point, Alsop was asking what each of us believed was important for the future, and Leonsis looked at me and jokingly said, "Who would care if you [a Time Warner media site] went down for a day?" His message: The e-mail service provided by AOL had become such an integral part of the business environment that its failure, rather than a mere disappointment to those seeking information or entertainment, was a major event. It was a short jump from regular e-mail use to daily Web visits. I knew then that the Web was itself about to become a central business tool.

The fourth watershed event occurred at home, when it became crystal clear to me that the Web was going to be an important player in the future of retailing. One day my seven-year-old daughter was baffled by why she couldn't find Pleasant Company's *American Girl* catalog (now a division of Mattel) online. Though the company has a great Web site, when she hit the button for the catalog, she kept receiving information about how to have it sent to her by mail. After twenty minutes of searching the Web site, she asked for my help. "I don't think it's online," I told her. "I think you have to send away for it or call the company."

"That's stupid" was her unequivocal reply, more blatant than a high-priced consultant but just as accurate. The younger generation has already reversed the order of things: To her, if you want something, you look for it by going online.

Ironically, in his best-selling book *Burn Rate*, Michael Wolff describes our efforts in launching Time Warner's Pathfinder (the first Web site from a major media company) as the single most important "event that got the business started, that precipitated the onslaught." Our launch was, according to Wolff, "the Internet equivalent of the assassination of Archduke Ferdinand," which led to World War I. For me, this was a major public event, but my thinking was shaped by the more private encounters discussed above.

HOW THE BOOK IS ORGANIZED

Part I of this book is my analysis of the rapid growth of the use of the Internet as a business medium and the implications of this growth for businesses of all sizes. Of necessity, this part of the book is a combination of three things: (1) a discussion of what is already happening, (2) a look at the dynamics that underlie this evolution to date, and (3) a discussion of the challenges for the business world based on this understanding of the underlying forces at work. Like all major changes, the widespread adoption of the Internet creates a combination of opportunities and threats for established businesses.

Part II of *HyperWars* is designed to help you successfully navigate the turbulent waters that are inherent in this period of rapid change. This part of the book is my analysis of what firms can, and in many cases must, do in order to prosper in the next phase of business as shaped by the growth of the Internet. To develop these ideas, I have drawn on a number of resources:

First, I have looked at what leading-edge companies are doing today. There is almost always something valuable to be gleaned by looking at specific examples of how successful companies—both large and small—are embracing the present and gearing up for the future.

Second, I have looked at existing management theory and its practical applications. My overriding focus in looking at prevailing management wisdom has been to hold it up to a new light, look at it through the lens of the Internet-contoured business world, and ask: "Do these ideas still provide businesspeople with valuable guidance? If so, do they need to be modified to lead to success in our emerging competitive environment?"

Third, using the analyses developed in the first part of the book, I have developed new, highly practical ideas about what businesses need to do today in order to survive and prosper tomorrow.

Part III of the book describes my personal view of the ongoing future development of the Internet. This perspective is included as an additional guide to assist you in the rapidly changing business climate.

Finally, Part IV, The HyperWars Audit, is designed as an easy-to-use straightforward means for you to evaluate the battle readiness of your company.

Here is what I want you to take away from *HyperWars:* a new and deeper understanding of how the Internet is fundamentally changing the nature of business and why, so that you'll be in a better position to think about how to make choices for your own business, whether it's large or small. It is also my goal that this book leave you with a number of practical ideas for ways you can improve your business immediately.

In college, I studied with a professor who constantly said, "Always remember that deciding to do nothing is also an active decision."

As Internet businesses have grown and branched out into multiple industries, my personal conviction is that in almost all situations, actions that promote changes are required if existing businesses are to survive and maintain their prosperity. I urge you not to decide to do nothing.

PART I

THE NEW COMPETITIVE REALITY

Chapter 1

The Emergence
of HyperWars

T his is not a book about the future. It is a book about what is happening now. In a short space of time, the World Wide Web as a tool for businesses has moved from an experimental start-up phase to an established part of the business environment.

All of this development has happened at an extraordinarily rapid pace. Only a short time ago, mainstream businesspeople were still asking whether the Internet would, in fact, turn out to be the CB radio of the 1990s. Today, we are more likely to agree that the Internet will, like the telephone, ultimately make its way into all offices, homes, and schools.

As Internet growth spreads, the way business is conducted is being profoundly altered. As a result, we have a new field of winners and losers. Formerly profitable businesses are obsolete, new classes of businesses compete with established entities, and successful businesses stay successful governed by a new set of rules.

Just as the telephone and the computer spawned entirely new business practices, the Internet is similarly transforming the way successful businesses must operate. Now executives at companies large and small are looking up from their spreadsheets and saying, "What happened to the great, profitable business I used to have? Why doesn't anyone need my products or services anymore?"

The pressing need to continually create innovative, better products to stay one step ahead of the competition defines business today, the world of HyperWars, where any business of any size in any industry must constantly be "at the ready" to do battle to maintain its turf.

The rise of the Internet has occurred at such a breakneck pace that few of us have had a chance to fully understand the magnitude of the change that has created the new world in which we now do business. It's worth noting that before the rise of the Internet, car buying through dealers fundamentally had not changed since the 1950s. Now there's a general belief that online car buying will account for 25 percent or more of all new auto purchases in the year 2000. *In effect, the Internet has brought about more change in the auto industry during the past three years than has occurred in the previous forty. For manufacturers and car dealers, an entirely new sales environment is emerging.* This is particularly significant both because automobiles represent a large chunk of our gross domestic product and because they have always had a strong symbolic meaning. At one time, it was said that "what's good for GM is good for America." In the space of a few years, GM and all other facets of the car industry have moved online. So, too, will the rest of the country.

Another way to put the rapid growth of the Internet into perspective is to compare it to other media. It took television thirteen years to reach 50 million domestic viewers, while it will take the Internet only four years to reach this number of people.

The reactions of several companies to the onset of HyperWars provide a snapshot of what is happening on a daily basis throughout today's economy:

The Brokerage Industry
Is Caught Asleep at the Switch

All brick-and-mortar companies that think "the Net won't hurt them" should listen to the story of how a virtual David (E*TRADE) has successfully challenged a mighty Goliath (the brokerage industry).

The entire history of change in the brokerage business has been shaped by the application of technology in order to lower costs (and thus prices) and to find ways to give the customer more control over his or her trading. In the eighties, discount brokers such as Schwab, Fidelity, and Quick & Reilly emerged at the expense of traditional brokers. They were successful because of their ability to provide low-cost trade executions. They proved that many customers preferred a lower commission, even if it meant they got less in terms of advice.

It was therefore *absolutely predictable* that with online stock quotes and growing consumer access to the Net, low-cost, Web-based brokerage services would develop. However, the giants in the industry didn't wake up to the new reality.

In the meantime, start-up competitors such as E*TRADE and Ameritrade entered the market and offered low-priced stock-trading opportunities to consumers. They gained market share, and as the Internet grew, traditional brick-and-mortar businesses began to realize they had made a significant mistake. Schwab, Fidelity, and a host of other established industry participants were forced to offer Internet-based trading with commissions cut to compete with those established by the online upstarts. This new, unbridled competition has led the value of the average online commission to drop by more than 50 percent in one year (from $47 to $21), and further decreases are likely. In fact, the competition is now so intense that respected analysts anticipate that some online brokerages will offer commissionless trades, with these companies drawing revenues from behind-the-scenes activities such as order flow, uninvested funds, and margin balances.

In the second round of this battle, Schwab used its resources to capture over 30 percent of the nation's online trades; yet, it is still locked in a fierce and expensive struggle. And competitive pressures forced Fidelity to change its online pricing structure three times in less than a year.

In addition, intense competition is an expensive game. The cost of acquiring new customers has soared. Brokerages can expect to acquire one new customer for every $300 spent on advertising and marketing.

For brokerages, Internet-based trading quickly became a critical component of the business. Today, a large and growing percentage of all stock trades by individuals are entered over the Internet—this is particularly shocking when you consider that only four years ago this segment didn't exist.

Had the traditional businesses been first, and had they been willing to adjust their rates according to the lower cost of doing business through the Internet, new entrants could never have gained a foothold.

The New York Times and the Newspaper Industry Proactively Defend Their Turf

Faster? Cheaper? Better? These are the questions each company must ask itself when evaluating whether or not its business is threatened by HyperWars.

At the dawn of the online age, most experts expected that one of the first virtual businesses would be online real estate ads. Why? Because reading ads online would be *faster, cheaper,* and *better* (because it's easier): Potential buyers could specify price range, community, number of bedrooms, special features (such as a pool), and instantly search databases of available listings. One or more pictures of each home could also be accessed; something not usually afforded by the paper.

Classified ads, including real estate, car, and job listings, account for one-third of newspaper revenues, so experts who noted the coming phenomenon of online classifieds also predicted the demise of the newspaper industry. Yet, the industry is healthy, and online listing companies have not supplanted newspapers as the primary source of classified information. Why? Because, unlike the brokerage business, the newspaper industry has behaved proactively. Major newspapers such as *The New York Times* recognized the threat early and began posting their real estate listings in consumer-friendly, searchable databases. By being first, *The New York Times* and other newspapers have held on to their franchise and have thus far blocked potential competitors from stealing it.

Several newspapers also had the foresight to realize that even broader change was necessary, and they've spent the past year or two reinventing the way they do business. Since the online world destroys geographic boundaries, several realized they might mutually profit by banding together. In order to hold on to revenues from job listings, six major newspapers, including *The Washington Post* and *The New York Times*, launched the Career Path site. This Web site now receives more than 1 million visitors a month and may well be a model for changes yet to come.

In fact, a consortium of newspapers is working to re-create the success of Career Path with Classified Ventures, an online listings firm designed to keep other classified revenues such as apartments, real estate, and automotive listings "in the newspaper fold" and to take advantage of newspapers' already existing relationships with local buyers.

That said, *The New York Times, The Washington Post,* and the rest of the industry must continue to keep looking over their shoulders. Although these companies may have won the first round, a raft of competitors, including AOL, Yahoo!, Realty.com (which is affiliated with the National Realtors Association), as well as the most powerful competitor of all—mighty Microsoft—have entered into the business of online real estate information.

To try to unseat the long-established dominant players is a costly challenge that would daunt all but the most robust, so this next round will be worth watching. Although the outcome can't be predicted, one matter can: The competition will be intense and the battles will be hard-fought.

Virtual Auto Services Reinvent Car Buying

You live in New Jersey but had a new car delivered to you from a Vermont dealer who sold it to you for less than you could have paid in your area—and you didn't have to negotiate with a salesman? Now *this* is the way to buy a car!

Industry polls have shown that car buyers want only a few simple things: convenient shopping, a good price for the car with the features they want, and the pleasure of not having to deal with a car salesman.

However, anyone who has bought a car by traditional means knows that it isn't easy to put those elements together: If car buyers want "convenience," they will often have to pay a higher price for the car because they talk only to their local dealer; if they want "best price," it will usually require the time-consuming process of going from dealer to dealer throughout a region and haggling over price with a pushy salesperson.

But now Auto-By-Tel and a number of similar services have turned the car industry on its ear by rethinking the way cars are sold and making the customer king. Auto-By-Tel operates as a virtual clearinghouse for a limited group of car dealers who have signed on with the company to do what they can to protect their turf. An online customer provides Auto-By-Tel with the specifications on the car he or she wants, and Auto-By-Tel essentially locates the dealer with the best price. If desired, Auto-By-Tel will also take care of loan arrangements and insurance, totally removing the "hassle" factor from car buying.

As a result of this online sales explosion, Auto-By-Tel has, to date, provided over 1.5 million consumers with quotes. Chrysler anticipates that 25 percent of all vehicle sales will come via the Internet by the year 2000, and I anticipate an even higher percentage.

However, stay tuned. This industry, too, will continue to change at a blistering pace. General Motors has been particularly active in developing innovative Web sites with the potential to play a central role in the selling process. Through its CarPoint site, Microsoft is also making a bid for this business.

At the same time, the growth of online car sales will inevitably have a significant impact on new car dealers, who are primarily local businesspeople. Many observers believe that the dealers' role in the sales process will change and that their numbers will inevitably shrink.

Whatever changes are coming, one thing is certain: The way people buy cars is going through a fundamental change; we can expect everything to shuffle around again, and only the nimble will survive.

Intuit Devises a New Strategy to Avoid Becoming a Casualty of the Net

Intuit discovered firsthand that owning a market niche (in this case, by selling the most popular personal finance software, Quicken) isn't necessarily valuable when new technology comes along. Already threatened by the software package Microsoft Money, the company actively addressed what *Fortune* magazine called "an even greater threat to Intuit . . . the explosive growth of the Internet."

New personal finance tools, developed by online content providers with specialties such as banking and real estate, are now being distributed piecemeal via the Net. As these products evolve, customers will pick and choose the pieces they need, and they'll no longer spend hours and hours entering figures; the new tools will have active links to market updates, bank account information, and other data. The proliferation of these new tools has the potential to make Quicken's shrink-wrapped software obsolete. These factors prompted *Fortune* in August of 1997 to question the company's future with an article entitled "Is Intuit Headed for a Meltdown?"

Aware of approaching calamity, Intuit set out to redefine its strategy. Building on the value of the Quicken brand name, the company decided to become a central hub for Internet financial activities of all types, and subsequently launched Quicken.com, which offers access to mortgages, insurance, small business loans, and other services. Its goals: (1) to become the place that consumers go when making important buying decisions related to financial products and (2) to then offer consumers choices among the best of the best. Intuit expects to create revenue by selling ads that surround the content of the site, and the company will take a fee for the business it feeds to financial institutions participating in its "mall."

Intuit was able to recognize the need to change its culture and adapt to an environment where speed was critical to success. As *Fortune* wrote in a follow-up article on Intuit just eight months later in April of 1998: "On the Web . . . you are either fast or last. There is no time for the intricacies of twelve-month product cycles and telephone-book-thick product specifications typical in the software industry."

To date, Intuit appears to have successfully steered a new course that embraces the Web. Nonetheless, the company is likely to face fierce competition on two fronts: As banks and brokerages move online, they are likely to expand the services they offer and similarly attempt to be the consumer's gateway. In addition (as detailed in chapter 2), the proliferation of product comparison sites and software agents that seek out products available on the Web will threaten Intuit's business model.

A STATE OF HYPERWARS

The examples above span a range of industries and involve different types of businesses. There are, however, certain common themes that illustrate the significance of what is happening in the business world today.

Fundamental change is the key reason we've entered a new era of business competition, that of *hyper*competition. The Net, which seamlessly connects businesses to business and businesses to individuals and individuals to individuals in a new way, has the power to wreak havoc on everything in its wake. And while major innovations such as electricity and the phone for the most part made the world easier, changes wrought by the Internet often force companies to prepare to do battle. Consider these developments, only a few of the many I'll point out to you throughout the book, but ones which I consider to be particularly significant:

Geographic Boundaries Are Becoming Irrelevant, and Virtual Businesses Are Threatening Smaller, Brick-and-Mortar Businesses

Using the Internet, both business and personal customers can now literally "shop the world" in their search for the best price.

Five years ago, if local retailers were asked to list their major competition, the local bookstore would mention great concern about the chain superstore being built in a nearby mall; the car dealer would tell you tales of how the dealer down the street snatches customers away from him; and the funeral director might scratch his head when asked about competition on casket sales, and then he might mention the one or two other funeral homes in town. The same was true of many financial services: Local banks were the primary sources of mortgages in each town, and local insurance agents competed among each other.

Many businesses have thrived in part because of geographic protection. Customers—even corporate ones—have always liked doing business face-to-face with people they know, and if a product is involved, the consumer is always happy if he or she can walk out with the exact item desired.

To a large extent, the Web shifts these dynamics, simply because it offers benefits in the form of *substantial* time and cost savings that often make customers willing to give up the face-to-face, personal interaction of doing business locally. Online businesses can take advantage of automation and other large economies of scale to provide offerings that local businesses can't match. Lean operations, the creative use of automated systems, the absence of costs associated with physical brick-and-mortar outlets, and economies of scale resulting from a single processing center make it possible for online companies to offer products that are often superior to those offered by local competitors. As far as the customer is concerned, if the Web can deliver the same product or service faster, better, or cheaper, then that's too good a buying opportunity to give up—thereby creating a serious threat to established local businesses.

These dynamics are at work in the mortgage industry. Until recently, mortgages were essentially a local business. Now, lower-cost national, online providers are actively competing in this arena. For example, the American Finance and Investment Company (AFI), a subsidiary of Virginia First Savings Bank, and other banks are transforming the home mortgage business by offering competitive loan rates to out-of-region customers and simplifying the application process: The customer fills out and submits an application online, which then triggers the electronic retrieval of the applicant's credit report; these are forwarded to a "decision engine" to quickly determine eligibility. The customer receives approval notification quickly, usually in less than five minutes.

Since the process is entirely automated, AFI has two advantages over traditional competitors: (1) Its cost to process a loan is far lower than banks that rely principally on loan officers. It is, therefore, typically able to offer better rates than local competitors. (2) The company is able to provide the customer with a convenience that was previously unknown. Just a short time ago, it would have been impossible to imagine that from one's own home, in under five minutes, one could obtain a new mortgage.

Speed Is More Important than Ever

Through a variety of mechanisms discussed later, the Internet accelerates almost all competitive dynamics, creating hypercompetition. Certainly one of the reasons for this acceleration is because online stores and product manufacturers can monitor what the competition is doing online. With this information comes a desire to counter, as quickly as possible, emerging competitive threats.

At the same time, the nature of the Internet makes it easy to introduce new products, services, or pricing plans at a moment's notice. This, too, leads to a faster-paced environment as companies square off against each other.

Finally, the Internet permits companies to save time on internal operations through a combination of seamless communications and automation. This, too, speeds up the nature of competition simply because companies that are using these new systems are more nimble than those that aren't.

New Competitors Are Emerging from Unexpected Places

In the past, a company's competitors were reasonably obvious. Now, in addition to traditional competitors, entirely new types of competitors are arriving on the scene. As shown in the above examples, these may be online-only companies. However, they may also be brick-and-mortar companies that are expanding their reach in new ways through the Internet.

In addition, as companies fend off one set of entrants into their markets, new entities—based on new ideas and new technologies—are making themselves known. As a consequence, companies need to be vigilant and on guard in a way that was not as critical in earlier eras—this, too, is a war mentality.

The Intensity of Competition Is Reaching a Level That Has Never Been Seen Before

While the Internet makes it inherently easy for businesses to alter strategy in response to their competitors, companies must remain alert and nimble in order to react quickly. It is literally possible for a firm to move from dramatic success and market leadership to "has been" status in a period of months. This is also the "hyper" aspect of HyperWars, and companies that cannot successfully adapt to the new competitive realities will not survive.

As a result of all these changes, especially the increased intensity of competition, the rapid development of new competitors, and the acceleration in the speed of change, companies are in a new state of battle. Each entity must be constantly thinking three steps ahead (and how to get there as quickly as possible), as well as watching for rear or flank attacks from known competitors and the opening of another "front" by new competitors.

CHANGE IS NECESSARY FOR SURVIVAL

To survive, many companies will need to change radically and adopt new ways of doing things, just as Intuit moved decisively to develop new businesses as it began to doubt Quicken's lasting strength. What is particularly noteworthy about this previously cited example is that this change also required that Intuit adopt an entirely new set of "operating values." As discussed earlier, in moving to the Web, Intuit made a giant cultural leap. It recognized that success on the Web required faster action than existed in the company's highly valued development process.

The intention of *HyperWars* is to provide businesspeople with the necessary insights to fight smart amid an ever-changing, ever-increasing group of competitors.

Chapter 2

How the Internet Changes the Competitive Battleground

Over the past several years, I have been a featured speaker at a wide number of industry gatherings. Not unexpectedly, the most common audience questions concern what the future will look like: "Will my business be helped or hurt?" "What should I do to survive?" Until recently, the pace of change has been so great that attempts at answering these questions were often unsatisfying.

As I have talked further with industry leaders, been active in creating several more Internet businesses, and written this book, I have become more and more certain that I can identify the elements that will play a central role in how the Internet economy evolves. I firmly believe these elements have the potential to greatly harm our economy or to power us into prosperity for the twenty-first century. The best way to discuss these ideas is to provide two contrasting visions of the future:

The Potential Downside of Internet Growth   More than any other aspect of the Internet, shopping "bots," software that searches the entire

Internet for products and services based on predefined criteria (price being the most popular), have the capacity to offer great benefit to consumers while potentially bringing great harm to businesses.

Today the major Internet portals, including Excite!, Yahoo!, and Infoseek, all have shopping initiatives with participating retailers that are, at least in part, powered by bots (also called "intelligent software shopping agents"). A visit to the BotSpot on the Web will reveal that there are already over 250 easy-to-use bots that are available for businesses and consumers. Each bot typically has a specialized function, ranging from financial bots that perform stock-related functions to news-gathering bots to auction-related bots.

The more businesses and consumers rely on price as the sole criterion for purchase, the more price pressure every manufacturer and retailer will feel. Bots that search for the best price have the potential to trigger ruinous price wars. At some point, price wars can ruin industries or set back their development by several years through lack of funds for investment in new and better services. To me, this is the frightening downside of a world where customers can buy from anywhere and essentially name their price.

The Internet is already demonstrating some of this behavior. The online computer retailer BUYCOMP, for example, is programmed to look at its competitors and then set its prices so that it will always come up as the lowest in any comparison-searching process. At the same time, we are witnessing deep discounting on the part of large Internet retailers to attract customers. For ongoing businesses, this type of discounting is unlikely to be sustainable.

Who'll suffer first? In this type of economy, small businesses will be the most vulnerable because they'll be faced with competing against inexpensive automated nationwide services that operate from a lower cost base. However, no business will be immune to destruction. If online prices become so low that businesses can't make money, the end scenario for a good number of Internet businesses as well as large brick-and-mortar companies is quite clear.

An Alternate Vision An alternate, also realistic, vision of the future is somewhat less forbidding. In this case, businesses learn how to effectively compete in a world of HyperWars. Businesses master the skills detailed in the second part of this book, and ruinous price wars are avoided. These skills include taking advantage of the many new capabilities offered by the Internet: for businesses to forge ever stronger relationships with their customers, for businesses to provide customers with expanded services and convenience, and for businesses to develop new types of sales initiatives. The strategies discussed in Part II of this book are designed to assist businesses in ensuring this version of the future.

Now, let's explore in greater depth the dynamics that underlie these ideas and how the competitive battlefield is shifting.

UNCHARTED TERRITORY

The pace of change in the past year has been so incredible that a growing population—from executives to retirees—feel they have more than passing familiarity with the Internet. However, this newly claimed "knowledge" of Web sites and portals and e-mail techniques belies the fact that the Internet is still so new that there are no rules for "play."

Anyone who has ever fantasized about being a gunslinger in the Wild West, where rules were made up as the community developed and disputes were settled by landgrab, brute force, and speed—not negotiation—can live that fantasy now. At no other time in recent history have we been on such uncertain ground, with territory demarcations constantly changing all around us. The World Wide Web has opened up a vast new world where anything can happen: The rules have changed; alliances are shifting; there are new ways to reach the consumer; and the "little guy" now faces off against bigger businesses and sometimes may be able to bring down an industry's most admired sharpshooter. (Upcoming "shoot-outs" will almost certainly involve price wars.)

WHAT HAPPENS WHEN "BEST PRICE"
TAKES MOMENTS TO FIND

Internet services, including comparison services, software agents such as bots, and new digital middlemen, enable easier access to information and, as a result, fundamentally change the competitive battleground.

The dedicated shopper has always taken the time to go from store to store or to make some calls, comparing prices and product offerings. Checking a relevant copy of *Consumer Reports* for feature comparisons would further enrich a consumer's knowledge.

Stores that pitted themselves against each other generally rose to the occasion and even employed "comparison shoppers," who cruised the aisles of competitors, reporting on pricing so that the store owner could price competitively. And signs touting "We'll Beat Anyone's Prices" covered any store that wanted to preserve business no matter what.

However, price comparisons of the past involved effort on the part of the buyer. At some point, many shoppers would say, "I know this retailer, and I know this is a pretty good buy. Finding a better deal might take me a week's effort to save a relatively small amount of money."

The Internet radically changes this dynamic. Not only can consumers move easily from one Web site to another, checking out the products and services at each, but many Internet services are specifically designed to promote comparisons among products or services within a category, and another growing class of businesses — digital age middlemen — are actively working to create bidding situations among several suppliers that lead to lower prices for consumers and businesses. These new classes of Web sites raise the stakes throughout an industry. Competition takes two forms: between manufacturers or service providers — for the product or service to be purchased — and between "stores" as the place to purchase.

Companies now operate in a world where comparison shopping is almost effortless, geographic boundaries are eliminated, and the pace of activity is far faster.

WHAT HAPPENS IN THE MARKETPLACE

In the online environment, anyone offering a particular product competes with everyone else offering that product or something similar. On any given day, a customer may be able to locate several—or several hundred—online stores selling what he or she wants, and the customer can then determine from which vendor to purchase the item. A recent search using a comparison service identified that a specific Epson printer was available from thirty-six different stores on the Internet at prices that ranged from $199 to $260.

In many ways, it's as if the consumer has gone shopping at Wal-Mart with representatives from Target, Kmart, and a host of local discount stores tagging along, shouting what their store will offer in terms of product and price, whenever the customer stops to browse. That's brutal competition!

Bots, the comparison services, and the new digital age middlemen can all work in a variety of ways. Some comparison services are operated by retailers; the site generally posts information on how specific products sold by the store compare in price and quality with the competition. Other services are stand-alone Web-based businesses offering comparisons among a variety of manufacturers.

Bots (Intelligent Software Shopping Agents)

As described at the beginning of the chapter, bots can be designed to search the Web for particular products and the available pricing and product reviews. One notable example is the service offered by Excite!, one of the large Web search engines. Excite!'s Product Finder, powered by Jango, is an easy-to-use shopping service that does not require the installation of any special software on one's own computer.

To search for a particular product, the user chooses the category, the type of item, the manufacturer and the model (if it's known by the user), and then clicks on "Find Prices" to see a list of the many Web stores carrying this item and the price at which it's being sold at each location. In a widely publicized upgrade, Product Finder was modified in mid-1998 to integrate online auctions and classified listings into its online shopping search.

Today Product Finder, along with hundreds of other "bot" services that are now on the Web, is still in its infancy. However, it has already stimulated the creation of sites that feature comparison shopping facts, and I guarantee that these bots will play a role in changing the future of commerce.

Sites Offering Price Comparison

On the Web, competition for customers is fierce. Because bots can shop the Web for a customer, companies have been forced to design their Web sites with comparison shopping in mind. The Net has facilitated explosive growth of comparison services that stretch across the full spectrum of products and services. It's impossible to even estimate how many exist today, and new services—for unserved categories or with better features in categories that already have such services—are constantly popping up:

"Side by Side" New England Circuit Sales (NECX), one of the largest retailers of computer equipment and software on the Web, has built a very specific comparison capability called "side by side." This capability allows the buyer to select up to ten items within a category (ten different models of laser printers, for example), and the NECX site will then display a lengthy list of the key features associated with each of these products in a side-by-side table. The prospective buyer can look across at the features or

combination of features that are important to him or her and see which product performs best on these dimensions.

This NECX service helps potential buyers decide which of many products carried by NECX they should purchase. It is both a valuable service and an example of a capability that can be offered on the Web and not (to nearly the same extent) in the physical world.

NECX is a company that realizes it is in a hypercompetitive world, and it was also among the first retailers to take the idea of comparison shopping to the next level: Once a customer has selected a specific product (out of a list of 1,000 best-selling computer products), NECX will show the customer a chart that displays both the price of the product at competing online stores and the percentage difference in price between NECX and each competitive site. Here, NECX's strategy is to be the starting point for the customer's shopping exploration. If competitors are offering the same products at essentially the same prices, NECX expects to get the business because the customer is already at its site.

NECX's service, which provides comparisons with other retailers, demonstrates how brutal competition for buyers has become. Competition has reached the point where businesses know consumers will be comparison shopping anyway, so many have adopted strategies to make it easy for the consumer, also illustrating that they stand behind their goods or services.

Dell has similarly recognized the inevitability of comparison shopping and configured its shopping experience accordingly. The Dell site will store a custom design of a computer system for two weeks; this feature implicitly makes it convenient for consumers to visit Dell, determine their appropriate buy, and then investigate the offerings of competitors.

Information Providers The unique attributes of the Web have also led to the creation of comparison services that don't sell any products. They act as "information providers" and often make money like the Yellow Pages—they typically provide listings of all companies, but the service then charges for advertising and sometimes collects a fee for a bolder or enhanced listing. However, no Yellow Pages offers the capabilities of these services or wields the potential to affect sales in such a powerful way. As noted above, there are hundreds of comparison sites—covering the wide range of goods bought by businesses and consumers. The following examples demonstrate how they work:

CNET Shopper.com (formerly ComputerESP.com) is a site specifically designed to help buyers find the absolute lowest available price on computer software or hardware. The consumer enters the name of the wanted product, and the site generates a list that shows the price of the item at each store and provides hot links to the particular store. The range of products offered and the depth of pricing information is almost impossible to grasp—you really need to see this to believe it. This site provides pricing on over 100,000 different software and hardware items sold on the Web. Moreover, the site updates approximately 500,000 prices each day, so it should catch special offers and sales on certain items. Certainly, nothing like this service was possible before the development of the Internet.

Travel services are booming on the Net, and Preview Travel, a reservation company, offers a service called Farefinder that posts the lowest available fares from major U.S. cities for specific days. In essence, the service compares prices across airlines for specific routes for specific days.

Wireless Dimension helps potential customers sort through the often dizzying array of cellular phone plans. Visitors specify the city they live in and answer a number of questions concerning the type of service they are seeking (range, amount of usage, etc.). The site then provides visitors with information on all of the offerings available in their area that meet these criteria. When more than one offering is appropriate, the site offers a valuable "side-by-side" feature that lets visitors compare the individual features of each plan.

Anyone looking for the best prices on mortgages would certainly want to check out BestRate. Visitors indicate their state of residence and the type of loan they are looking for, including whether or not they want to pay upfront points with the loan. The site then generates a summary of the lenders offering the best mortgage rates that meet these criteria, the rates offered, with a link directly to the sites of the offering lenders. Then these lenders typically allow consumers to complete all or a portion of their applications online. The convenience is extraordinary, and the entire process takes under two minutes.

"What's my best buy if I want to spend $300 on a television?" is the type of question customers can ask at NetMarket, Cendant's Web site that offers prospective buyers the opportunity to search for specific products with specific features at specific prices, from their database of 800,000 brand-name items. Products range in categories from electronics to home appliances and home furnishings. Anyone can shop at NetMarket; it is a buyer's club and members also receive special benefits, including low-price guarantees.

The net effect of the explosive growth in comparison services is twofold. First, it places incredible pricing pressure on Internet merchants or service providers. To attract a customer, a company may not have to offer the absolute best price on an item, but it does need to be competitive. And merchants must realize that at some point consumer loyalty to a particular store or supplier weakens in the face of significantly lower costs for the exact same services or products. Second, comparison services place similar pressure on the manufacturers or providers of products. Buyers now have access to more information than ever before on which products provide the best features at the best prices.

Both of the above factors contribute to the creation of a Hyper-Wars environment; yet, this is just the beginning. New digital age middlemen only intensify the competition.

Digital Age Middlemen

An additional factor in the spread of HyperWars is the emergence of a new kind of business, "digital age middlemen." Specific examples of this kind of business range from consumer services like the Lending Tree (a mortgage service that arranges for different lending institutions to compete for the borrower's business) to services such as FreeMarkets Online (a bidding service for companies seeking industrial supplies). This broad range of entities may share one or more of the following characteristics:

First, they create a "more perfect market" for buyers and sellers by improving the information available to both sides regarding the demand and supply of products.

Second, they generally serve as electronic gathering places where buyers meet potential sellers for specific industries or types of products. In this sense, successful digital age middlemen are typically vertically focused and aim to serve a very specific market, such as buyers of specific financial products, or specific kinds of engineers or electronics firms. Occasionally this leads to selling the same products at different prices. Prior to the launch of Priceline.com (a service that lets buyers name their own price for a variety of items such as airline tickets), company founder Jay Walker and I discussed his then-confidential, upcoming innovation. Classic microeconomic theory says that different buyers will pay different amounts for the same product because of the different value this item has for each buyer. Walker told me that with Priceline "we are going to be bringing the real world much closer" to this vision of microeconomics.

Third, the digital age middlemen defy traditional boundaries. Location, company size, the time required to get competing bids from many potential suppliers, and the time involved in searching for potential new suppliers are all factors that have traditionally created "competitive advantages" for specific companies in specific situations; these services work to eliminate these constraints.

And fourth, they all provide a buying mechanism of some type. Often, digital age middlemen pit potential sellers against each other, and generally the one with the lowest price earns the business.

For buyers, the effect of digital age middlemen is to make goods and services available at lower prices. For example, FreeMarkets Online works for buyers by finding potential suppliers of industrial products to meet a specific RFQ (request for quote) and then holds an online real-time auction among prequalified suppliers. FreeMarkets says that the typical customer saves 17 percent in procurement costs using this service. Some customers have reported savings of 20 to 30 percent.

Glen Meakam, the founder of FreeMarkets, believes his company succeeds by filling an information void. Meakam has demonstrated that large industrial buyers are not always aware of the range of small and large suppliers that might meet their needs. "We focus on the buyers, giving them better information and pulling suppliers to them," he says.

As demonstrated by FreeMarkets Online, another aspect of these types of services, particularly in a business-to-business context, is that they open the door to new suppliers. GE TradeWeb, which is discussed later, similarly helps smaller firms access opportunities to reach new customers.

For buyers, of course, the downward pressure on prices is a welcome benefit of the Web. For sellers, however, digital age middlemen are a central factor in the emergence of HyperWars. By creating a "more perfect market" with full information, fewer geographic boundaries, and more potential suppliers, the level of price competition reaches a new degree of intensity, making it harder to generate profitable revenues. As these online middlemen proliferate, each transaction will feel to the seller even more like a battle in a war that is waged in hyperspace.

DIRECT SALES ARE EXPANDING, TRANSFORMING THE MARKETPLACE

Experience to date strongly suggests that the Internet will accelerate the shift to direct sales by manufacturers, thereby creating a major change in the way both businesses and consumers shop.

Over time, the percentage of all products sold directly from manufacturers or service providers to end users has been increasing, and the cost of having a middleman is being eliminated. This phenomenon has accelerated whenever the development of new technologies or new media has provided ways to overcome what are generally perceived as the four barriers to direct sales:

1. The consumer's need to see the item being purchased
2. Product complexity
3. Access to product information
4. The consumer's need to judge legitimacy. Is this person and company reliable or am I being sold snake oil?

For the most part, the Internet has quickly done away with each one of these obstacles and offers added benefits as well:

The Importance of the Visual Certainly a primary Internet breakthrough for direct selling is its visual capabilities. Unlike the telephone, the Internet makes it possible for consumers to see items—such as photographs of travel resort destinations, electronics equipment, or articles of clothing. The basic need of people to "see what they are getting" is the first hurdle the Internet surmounts to make it more compelling to sell items directly.

Making the Complex Comprehensible Over the years, I've worked to perfect a theory of how changes in media capabilities affect direct marketing. Here's what I now believe:

As a general rule, items that are easy to understand are sold directly to the end user, while items with complex or hard-to-understand benefits are sold through agents or other middlemen.

An example of this phenomenon is life insurance: On the one hand, term life insurance is easy to understand (I pay $100 for $50,000 of coverage), and so it is typically sold through direct mail and other direct mechanisms. On the other hand, whole life insurance is a complex combination of life insurance benefits, and so it is generally sold through knowledgeable agents.

When a medium can fully explain or show how a product works, then a large portion of the product's sales typically shifts from agents to direct sales by manufacturers or suppliers.

The Internet has a unique capability (interactivity combined with a low-cost avenue for communications) to make complex items such as financial products comprehensible. Because of the Internet's ability to filter, the consumer is presented with information directly relevant to his or her situation and therefore avoids wading through a stack of folders and brochures. A well-designed site can explain all of the product features and benefits of whole life insurance policies and annuities. As a result, these products can now be sold directly by their providers over the Web, since agents are no longer needed to explain the products.

Time and Self-Generated Recommendations In addition to making the complex clear and comprehensible, the Web saves customers time. The customer needn't answer any salesperson's prodding questions, and he or she can scroll through the screens as many times as it takes to fully understand a product. For example, a financial services company's site can also include interactive planning tools that help the customer to determine how an annuity fits into his or her overall financial plan and what size is best. In essence, this represents a system that employs interactive tools to both explain the product and recommend what consumers should purchase.

Convenience The convenience offered by the Internet is also a factor that will further the shift to direct sales. Everyone with Internet access can now shop for products and services at any time that's convenient without prior research or preparation. In the past, I might have wanted to buy a computer or financial product direct; however, to make an intelligent purchase, I would need to have preordered a catalog or brochure. All of this information is now available on the Internet at any time of day or night. As a result, the "preparation impediment" to direct shopping is removed.

Legitimacy and Consumer Trust As trusted consumer brands, such as L.L. Bean, have established active sales on the Web, the medium has become more legitimate. Moreover, as brands built solely in cyberspace have become recognized and trusted, consumer resistance to online shopping has lessened. Consumers hear their neighbors talking of buying through the Internet without getting "ripped off" and the "Can I trust you?" question is answered.

THE ULTIMATE QUESTION

Does the Internet toll the death knell for physical stores, or the need for "real people" (insurance or real estate agents, for example) to help consumers make decisions? Not quite yet. The need for sound (human) judgment, the social experience of shopping, and ingrained habits all suggest that the retail world will not suddenly become a ghost town. However, the fact that there is now a new sales channel that has never before existed creates an indisputable challenge to traditional businesses.

In my speeches and debates with industry gurus, I originally adopted a moderate view—suggesting that sales from brick-and-mortar outlets will only decrease at the margin. However, as the pace of change has accelerated even further and as we can see the outlines of the future of electronic shopping, my predictions have become less conservative. I now anticipate a major shakeout among retailers. *The brick-and-mortar shopping world will not go away, but the number of viable companies in it will shrink significantly.*

I believe that three types of retail businesses in particular will be affected: First, businesses that provide a limited social experience are at risk. If most consumers don't enjoy a particular shopping task, such as buying pool-cleaning supplies, then brick-and-mortar businesses that specialize in that area are likely to be vulnerable to online shopping.

Second, I discussed earlier the particular vulnerability of smaller, local businesses. These businesses have long struggled to survive in a world dominated by large chains. Unfortunately, low-price shopping on the Internet adds yet another reason they may have difficulty remaining viable.

Third, large chains with inefficient operations or those saddled with the high overhead costs related to maintaining a brick-and-mortar business will face price pressure that could well endanger their ongoing operations.

(In Part II of this book, we discuss ways for all types of businesses—including those based in brick and mortar—to survive. Don't have any "Close Out" sales yet.)

CHANGING THE SALES CHAIN: CHANNEL CONFLICT

Historically, consumer product businesses have operated with a well-defined chain of sales: Suppliers sell parts and materials to manufacturers, who sell finished goods to retailers, who sell to customers. However, in cases where the manufacturer is able to sell directly to the end consumer, the product costs are often significantly lower. This may reflect the following factors: (1) a middleman, who may add costs and not much value, is taken out of the process, and (2) the manufacturer has greater control of his or her inventory and the ability to introduce new products faster, since no products are caught in the supply chain.

Making the transition from selling through a middleman to selling direct (channel conflict) is a troubling issue many businesses are wrestling with right now. On the one hand, retailers naturally become "unhappy" with manufacturers who, in their view, go around them and directly to the consumer with a lower price. On the other hand, the manufacturer may observe cost-advantaged direct sellers who, because of lower prices as a result of lower distribution costs, are winning business and market share. In addition, a company may have its entire sales force dedicated to selling through the retail channel. How will the sales force feel (and react) if their own company starts to compete with them by selling directly over the Web to consumers? For the manufacturer, this is the dilemma of channel conflict: Do I stick with stores alone, or can I successfully implement a mixed strategy (stores and direct), which may ultimately lead to a significant or even majority share of sales through the direct channel?

Right now channel conflict is probably the most difficult, vexing problem faced by companies in this hypercompetitive environment. I repeatedly find that in discussions with executives at all types of companies this issue is at the forefront of their concerns.

The problem of channel conflict is a central reason the PC business is an industry on a high wire. If everyone in an industry moves through retailers, there is a fairly defined set of competitive rules. However, it is absolutely predictable that if direct selling makes sense, then someone will enter the market. It may even be a brash upstart with no roots in the retail distribution business. For the PC industry, that upstart was Dell.

In the second half of the 1990s, when PC owners compared notes about where they had purchased their computers and what they had paid, "I ordered it direct from Dell" became a frequently heard report, and before long Compaq and IBM realized they had a problem. PC orders for the two were failing to grow proportionately. Dell, the upstart computer manufacturer, was rapidly gaining market share. Even worse, Dell's prices were well below those of the two largest competitors, yet its profit margins were substantial. What was Dell doing right, that these other manufacturers had missed?

The increasing popularity of the Internet meshed perfectly with Dell's sales model of selling computers direct to customers. On the Web, customers could visit Dell's site, decide what they wanted, see images (something that wasn't possible over the phone), and place a direct order. Since Dell had no inventory tied up in a retail sales channel, the company had an estimated cost advantage of 10 to 15 percent over its competitors who sold through retail stores.

Moreover, Dell could react instantly to changes in consumer-buying patterns.

Before they could take on Dell, Compaq and IBM faced a fundamental dilemma—channel conflict. If these two giants started selling direct to consumers, they would disrupt traditional sales channels and risk angering their retailers, partners who accounted for their existing sales to consumers.

Dell's lower-cost Internet-oriented methods created chaos throughout the PC industry—chaos that continues to this day. Using their own brands, retailers now actively compete with everyone, including their suppliers. While Compaq and IBM both developed innovative techniques that permitted them to lower costs and deliver custom PCs through retailers, Dell continues to gain market share and grow in profitability. And IBM has recently opted to go around its distributors for sales of computer servers to its business customers in order to battle low-priced rivals, so further change is in the wind.

The importance of this issue cannot be overstated. And we have begun to see that some companies will "bite the bullet" and risk retailer retaliation in order to begin building a direct sales channel. For example, in June 1998, Hasbro announced that it intends to sell direct to the consumer, thereby adopting a mixed distribution strategy that, in part, circumvents its retail partners. This is just the beginning.

Keep in mind that if existing companies don't grasp opportunities, others will. HomeCom's InsureRate is selling annuities directly over the Web. HomeCom, an Internet applications company, had no previous experience in the insurance business; it entered the business to offer consumers state-of-the-art tools that would give them up-to-date information on buying insurance, then InsureRate would offer them cost options that were "consumer friendly."

It may take years for an industry to be turned on its head by direct sales, but product manufacturers need to plan for that eventuality. For example, one could argue that record labels would never sell directly online—a move that would effectively go around online music stores, not to mention brick-and-mortar music stores. But as direct distribution becomes possible over the Net and sites can be readily found using bots, it's hard to imagine that music sales won't one day be sold and distributed directly from record label sites and downloaded to consumers' computers. It is even possible to imagine that an artist—who may be more of a recognizable brand—will sell music directly from his or her own site. Never assume that lower-cost direct distribution won't happen because of channel conflict.

TWO GUARANTEED BENEFITS THAT ARE CHANGING COMPANIES INTERNALLY: INSTANT COMMUNICATIONS AND LOWER COMPANY COSTS

By seamlessly linking people and entities around the world, the Internet can accelerate the speed of new product development and introduction while lowering the cost of operations throughout the enterprise.

The Web offers a whole new way of setting up a business. If a company were never to sell a single "widget" online, it would still be worth mounting a Web strategy.

The Internet permits companies to cut costs through seamless, automated communications both within the company (intranets) and with the company's suppliers (extranets). Moreover, these communication systems can dramatically lessen the time involved to produce specific products because various departments and outside suppliers can communicate automatically.

Strategies 2, "Speed Is Everything," and 3, "Cut Costs and Increase Efficiency Using the Web," detail many of the ways companies are benefiting by using the Web. As companies are striving to compete in the emerging environment, they must also grapple with competitors who have put these new technologies to work: for faster and lower cost operations, for increasing the speed of their product development and deployment, and to more effectively manage a worldwide sales force at lower costs.

FOUR MORE REASONS WHY THE INTERNET UPSETS THE APPLECART

The Internet is creating fundamental widespread changes in the world of business. Comparison shopping, the breakdown of geographic boundaries, the increase in direct sales, the reduction in costs, and the incessant need for speed—all as a result of the Internet—have led to vast changes. But there are still more ways in which the Internet is affecting businesses of all types. At this point, your industry may be affected by one or all of the changes discussed in this chapter; very few businesses aren't affected at all, and if you're in this minute minority, prepare anyway. The tsunami is coming.

1. The Internet facilitates the sale of personalized products.

Consumers' desire for customized products has been well documented over the past several years. Who wouldn't want something made exactly for them? Similarly, businesses have always needed custom configurations—which were presented in paper-based orders. In recent years, the ability to manufacture mass-customized products, at acceptable costs, has been developed. The Internet provides the missing piece: the perfect sales channel.

The very nature of the Web makes it an ideal mechanism for custom-based orders. Buyers—whether consumers or businesses—can see pictures or illustrations of the components via the Internet, and the order specifications can often be tied directly to an automated system that checks accuracy, with the site saying, in effect, "Yes, feature A can be added to feature B," so the possibility of paper-based errors is reduced. As we will discuss later, Cisco found that its error rate on orders went from approximately 25 percent on paper-based orders to less than 0.1 percent for Web-based sales (using an intelligent configurator) that replaced other channels. This contributed to savings of hundreds of millions of dollars.

2. The Internet turns "free" into a strategic weapon.

Discussions of the Internet are often punctuated with phrases such as "But everything on the Net is free" or "I can find anything I want for free on the Net." As discussed in Strategy 8, companies do provide a great deal of value for free on the Internet. How to effectively harness the use of "free," or react to a competitor's strategic use of "free," while still running a profitable business is a new competitive challenge for most companies.

There is an additional dimension to this problem. In a Hyper-Wars environment, how does a company react when a competitor makes strategic use of "free"? The competitor may have a brilliant plan for using widespread distribution to generate alternate revenue streams, or the competitor may be making a last-ditch effort to survive. In either case, this is a central issue for companies that want to actually market comparable products for a fair price.

3. Delivery via the Internet is transforming several industries; it will soon affect others.

If you can log on to your computer, order a product, and receive it while still sitting at your computer, that is indisputably "faster and better." And because there are no delivery charges, the product cost is almost certainly "cheaper" as well.

Today software is routinely being delivered online; sometimes customers don't even need to order it. Your computer tells the seller's computer that an update is needed, and, presto, new software is installed automatically. This evolving distribution system will inevitably change the competitive dynamics of the software industry. For example, it is likely that software packages, which are always the latest release, will be available for monthly lease as opposed to the current "buy it and own it" model.

Elsewhere, I address a likely, similar transformation in the sale of music.

4. Products that can be sold through searchable databases (any type of offering where the buyer provides specifications and the seller tries to match those requirements) are going to become highly competitive online products.

Consider planning for your summer vacation. If you're scanning the newspaper for a summer rental in the general vicinity of your city, you might have to read the listings for several different geographic areas and then scan through the ads as grouped by a Realtor. Contrast that to searching for a summer home online: Log on, go to one of the classified ad sites (many, but not all, of which have been put up by newspapers), and then enter the qualities you're looking for: preferred township, timing, cost, number of bedrooms, and "walking distance of beach," for example. Within seconds a range of selections will be presented.

These offerings are unquestionably faster and better than working through traditional newspaper ads. They may even eventually cause some offerings to be cheaper, since you'll have all the listings in front of you and you can compare carefully. As newspapers fight to retain their revenues against Web-based competitors of all kinds—for everything from home purchases and apartment rentals to cars and job listings—new types of competitive behavior will inevitably result.

SPEED AND THE LAW OF TWOS

Al Reis and Jack Trout (*The Twenty-Two Immutable Laws of Marketing*) have persuasively argued that categories of businesses start as "a ladder of many rungs," but that all categories gradually become "a two-rung affair," where competition resides primarily among the top two contenders, such as Pepsi and Coke, Hertz and Avis, McDonald's and Burger King. It's not that there are no other competitors, but that the lion's share of the market ends up split among the top two competitors.

The Internet, however, accelerates the playing out of the law of twos. There is nothing gradual about what happens here. Because of its focus on comparative shopping, instant communications, and easy distribution of information, the Internet accelerates the speed of competition. As might be expected, the law of twos occurs even faster. This is clearly demonstrated in Internet-based businesses:

At the "dawn" of the Internet age, there were over ten browser companies. Now there are essentially two—Netscape and Microsoft. Similarly, all of the other major Internet business categories (thus far, music, books, auto buying, online stock trading, and travel) have consolidated rapidly, leaving just a few major players.

In fact, Bertelsmann AG, the German media conglomerate, chose to join forces with Barnes & Noble in the highly competitive online book sales arena. Instead of establishing its own U.S.-oriented initiative, Bertelsmann paid $200 million for a 50 percent stake in barnesandnoble.com, in what is generally viewed as an effort by the two companies to challenge the market leadership of Amazon.com.

At first blush, it may seem as though the idea of rapid industry consolidation and the law of twos is in conflict with a central theme of this book—that market success can be fleeting. In fact, they are complementary. The law of twos suggests that there will be consolidation *in an industry as we currently define it.* However, what I am also suggesting is that *the Internet makes it possible for new entrants to redefine an industry* (Internet-based stock trading, for example, redefined the discount brokerage business; online music and book sales sites are redefining retailing for these products; etc.). *When this redefinition takes place, the entire cycle starts once again: Typically there are multiple entrants pursuing this new idea of the industry and rapid consolidation then occurs.*

Although there has already been consolidation among many online businesses, the potential for radical change remains: Entirely new business concepts are likely to evolve and start the process of intense competition all over again. (Read in Strategy 11 about how you must never stop looking over your shoulder.)

HUB AND SPOKES
AS A NEW BUSINESS STRUCTURE

Natural market forces that lead to consolidation and the law of twos happen even faster on the Internet. The "best" products are anointed, information is rapidly distributed, and customers (who face no geographic limits) seek out these "hot" places. Moreover, as brands become familiar, consumers increasingly seek them out because they have credibility.

Unique to the Web is something I call a "hub-and-spoke strategy" that also facilitates the law of twos: A hub-and-spoke strategy is the increasingly common arrangement where a major retailer establishes relationships with other Web sites, promising them a percentage of any sales (generally 5 to 15 percent) that result from any traffic sent to the retailer by these "affiliates." It's not uncommon for a major affiliate program to have over 10,000 participating members, all of whom refer traffic to the central site. What's even more interesting is the extent to which this type of business model is proliferating: Refer-it is a site that lists credible affiliate programs for prospective members. As we go to press, it is approaching 500 distinct plans.

The Internet is already taking its place as a central tool in everyday business life. Each aspect of commerce is changing in light of the new capabilities available. The coming changes are so powerful that it is both terrifying and exciting. The first phase of Internet business has principally focused on the creation of new types of businesses: The next phase will involve the impact of these developments on brick-and-mortar businesses and on the economy as a whole.

The new capabilities that the Internet makes available for business-to-business commerce and for business-to-consumer sales will transform many industries from top to bottom. Winning companies will understand that in the emerging era success now requires new rules for action and the mastery by the organization of different types of skills.

PART II

ELEVEN STRATEGIES FOR THRIVING IN A HYPERCOMPETITIVE WORLD

For all types of businesses, the Web presents both threats and opportunities. In Part I, we stressed how the Web will inevitably create new and difficult competitive pressures. This section examines the flip side of this dynamic by introducing ways all types of businesses can harness the capabilities of the Web, not only to survive but to prosper in the new era of online competition.

Use the Internet as the World's Most Sophisticated Telephone

In exploring strategies for success in the developing environment, it is essential to recognize a fact that is often overlooked: The Internet is fundamentally a new communications vehicle. As a consequence, a large part of its value arises because it permits cost-effective communications—down the street or on a worldwide basis—that were not possible before its emergence.

Why is this so important? Because many people have a very different view of the World Wide Web. They will suggest that the Web is an entertainment medium—something that has more in common with the television than the telephone. This focus is easy to appreciate; the typical person is more interested in the new offerings on the Web that can entertain him or her than the less exciting details of enhanced communications capabilities. In addition, Internet use is the first activity in over forty years that has been clearly documented as something that causes people to spend less time watching television. It's therefore natural to think of it as a substitute for this medium.

I find it far more helpful to think of the Internet as an improvement on the telephone than as an improvement on the television.

Think back for a moment to the pre–Alexander Graham Bell days of dusty streets, general stores, and telegraph offices, and consider how the invention of the telephone started a seismic shift for commerce. Starting in the 1880s, businesses and customers began to benefit from the following changes brought about by Bell's invention:

Convenience. No longer was it necessary for a customer to visit an establishment, write a letter, or send a telegram in order to conduct business.

Interactivity. Because of the give-and-take of the telephone, matters could be ironed out almost as readily as if the business transaction were being carried out face-to-face.

Speed of Response. Business owners could answer questions and carry out orders more rapidly.

Now, the Internet takes these benefits and adds to them enhancements that catapult cost-effective communications capabilities to an entirely new level:

Product Visibility. If customers want to order something by fax or telephone, they aren't necessarily seeing what they order. On the Internet, customers can increasingly see detailed, colorful depictions of products and read extensively about a product's quality and components.

No "Help Wanted." While voice mail and auto-answering systems have reduced the need for human staffing of telephone systems, these procedures have by no means eliminated it. With the Internet, complete transactions can be carried out at any time of day or night, often with no personnel (and related costs) involved.

Availability Completely under Customer Control. With the Internet, visitors—potential customers—come to Web sites at *their* convenience, making them far more receptive to what companies have to say because the customers aren't being intruded upon (as happens with telemarketing).

One-to-Many Communications Performed Seamlessly. The Internet offers a one-to-many communications systems without losing the privacy or interaction possible by phone. A single posting at a Web site reaches as many people as visit the site that day.

Reduced Effort, Time, and Cost. The Web makes things easy and affordable.

Not all businesses are currently bringing in added profit via the Web yet; nonetheless, every business needs to be working on it in order to be competitive in the twenty-first century. This chapter outlines many easily implemented benefits, most of which can and should be put to use immediately.

TIME-SAVING, COST-EFFECTIVE SUPPORT FOR STAFF AND SALES PERSONNEL

The Web makes it possible for companies both large and small to develop new communications processes that save time and money while enabling faster responses to customer needs.

Many industries rely on widely distributed field sales forces that may consist of independent agents or company employees. In today's fast-moving business environment, providing these front-line soldiers with the most up-to-the-minute information and the best possible tools and support is critical to success, and by using the Web, companies can do so at far lower cost.

The insurance industry is one industry that depends on its nation-wide network of agents having current information about its products. In mid-1998, Prudential made a decision to invest $100 million in outfitting its roughly 12,000 field agents with notebook computers that provide them with an online link to the company. In an earlier six-month pilot test involving 500 agents, Prudential found that the number of life insurance policies sold more than doubled, turnaround time decreased from weeks to days, and commissions increased approximately 150 percent. Now by making the most current forms and brochures available electronically, the company anticipates that agents will be able to make sales faster, conclude purchases in days rather than weeks, and file all forms electronically.

For any company, no matter how large, a $100 million investment is a strong statement about its belief in the potential of a new communications capability to increase sales productivity.

Other companies employing the Internet for staff support range from JCPenney, linking its over 1,200 stores, to Baltimore-based Polk Audio, a sound system sales organization with 500 distributors worldwide.

Polk Audio has found that it can service its distributors far more effectively via an Internet connection. Formerly, if a distributor was out of stock on a particular speaker, the store salesperson would have to look up the price and availability, communicate with the customer, and then call Polk to order one. Delivery then needed to be scheduled.

Now a store salesperson need only check the information on Polk's Web site, and once an order is placed via the Web, the entire process can be tracked online, thereby resulting in time savings for both the salesperson and Polk headquarters personnel.

Businesses are finding that using the Internet to give company representatives easier access to information reduces the time it takes to serve customers and frees employees to focus on more productive tasks. As I will discuss later, they are also finding that processes that are automated from the start, such as the agent's submitting an electronic insurance application, provide an additional important benefit: dramatically reduced costs because of fewer processing errors and faster turnaround.

A SYSTEM OF VIRTUALLY
COSTLESS COMMUNICATIONS

The Web makes it possible to communicate regularly with a large volume of customers at virtually no cost.

Businesses can generally benefit by disseminating information; yet up to now, there has not been a cost-effective, satisfactory way of timely customer notification. Not only is direct mail costly (at a minimum of $500 per 1,000 mailing pieces), but the timing of delivery is erratic and an overwhelming amount of it is never even opened. The telephone is timely, but information disseminated by telephone is also costly and runs the risk of alienating customers who don't want to be bothered by solicitors.

Enter the Internet. The World Wide Web gives companies a low-cost method to communicate with existing customers and to reach out to potential ones with a timeliness that has never before existed.

American Airlines has established an e-mail program in which at least once a week the airline sends over 1 million of its customers time-sensitive e-mail notification of flights that are available at special reduced rates for each upcoming weekend. The benefit to the customer is financial savings on flights booked; the benefit to American is filling seats that might otherwise go empty, while constantly reinforcing the message "Fly American."

Over the years, airlines have tried several different ways of filling these seats at discount rates. None of these earlier efforts were particularly successful. As a communications capability, the Internet helped to solve what had previously been an unsolvable business issue. This is one example that shows how the effective use of e-mail can create a valuable new revenue stream for companies.

Fuel distributor Grenley-Stewart Resources, Inc., knows that trucking companies can realize big savings through pinching diesel-fuel pennies, so it uses the Internet to distribute information to truckers to help them minimize fuel costs. (Prices can vary considerably from location to location.) Currently the company posts contract diesel prices on its Web site, permitting fleet managers to help truckers map out routes and plan their stops based on prices. With the onset of this initiative, the company witnessed a 300 percent climb in sales in an eighteen-month period.

The new capabilities created by the Internet far exceed what could be accomplished with the telephone. Consider how a well-designed Internet customer-communications system can work:

1. Orders are confirmed by e-mail—first immediately after they are placed, and again when they are shipped out. The shipping confirmation notice includes an internal tracking number to help customers locate the package if it fails to arrive on a timely basis.
2. Customers can register for e-mail notifications of various kinds. By filling out an online form, customers can request to be notified about newly available products that are likely to be of interest to them.
3. "Missing" customers can be inexpensively lured back: If a frequent customer has not made a purchase for some time, the electronic retailer can send a $5 or $10 digital coupon to encourage a return purchase. These types of ongoing efforts to build loyalty can be triggered by well-designed automated databases, combined with virtually costless e-mail, to create an inexpensive, potentially high-return, customer loyalty program.

This suggests a central strategy for any business today: Gather e-mail addresses from customers (and permission to contact them using these addresses), even if you don't yet have an interactive Web site. Every business from a major manufacturer to a regional discount store to the local plumber will find that well-designed e-mail messages can be a low-cost, highly effective means of building profitable revenues. In Strategy 7, I discuss how a local pest-control business might benefit tremendously from a e-mail–based initiative.

ROUND-THE-CLOCK AVAILABILITY

Like a good catalog and 800 number, the Internet makes your company accessible to customers worldwide twenty-four hours a day. However, I like to say that "the Web is better than the world's greatest catalog." Here's why:

Additional visuals as well as more written detail. Catalogs face an inherent limitation: Paper and postage are costly. As a result, details—other views of a product as well as more written description—often have to be left out. So while the 800-number operator can read to customers the special washing instructions, if the product is offered on the Internet, the consumer can read the special washing instructions for him- or herself, scroll through a more lengthy product description, and in all likelihood, see more than one view of the item. One day (probably sooner than we expect) there will also be a way for customers to "see" themselves in apparel items (via scanning).

Expanded offerings. Today catalogs typically list only a portion of a company's offerings, simply because more listings mean expansion of printing and mailing costs. The Web obliterates this limitation.

At Time Warner, I was a strong advocate of the Book-of-the-Month Club's Web-based initiative. My reasoning: Those of the 1 million or so Book-of-the-Month Club members who are online definitely receive an improved version of the BOMC experience. Because of the cost of printing and paper, members generally receive a print catalog listing approximately 30 new titles and 200 past offerings. Online, customers can choose from 3,200 selections; there are also longer descriptions of each book, author biographies, and editors' comments as well.

BOMC demonstrates how the Internet changes the economics for companies with large inventories: Previously it was overly costly to communicate the breadth of offerings available to consumers. Now, after a site is created, it is almost costless.

Even infomercial-style television programs have their limits when compared with the Internet. While QVC's online store is profitably capitalizing on its existing cable TV–brand recognition, the company has found that on the Web it can offer even more variety. QVC Cable may demonstrate and sell a KitchenAid mixer, but the channel's Web site, iQVC, can profitably cross-sell a full line of blades, bowls, and attachments with the mixer.

Interactivity. Preview Travel, Inc., is now wooing business travelers by creating an interactive guide to destinations, based on data from Fodor's Travel Publications that goes far beyond any brochure or book. The company's Business Travel Center features a news wire that carries a constant stream of information on travel specials. It also allows users to scan hotels and restaurants and then print out the information as a personal mini-guide to take along. Preview can also book airline, hotel, or car-rental reservations using special discount arrangements with its partners. It would almost be impossible to consider this kind of at-home, on-demand customization if these travel packages had to be sold by direct mail or phone.

Using the Internet, the two owners of Kauai Exotix, an exotic flower farm in Hawaii that ships premium tropical blossoms direct to customers, have found that their twenty-four-hour availability on the Internet lets them surmount a central business hurdle. Because the Hawaiian Islands have a six-hour time difference to parts of the United States, the two businesswomen knew from their start in 1994 that they required a communications system where the business could function efficiently despite the time change. What they've found is that the time difference actually gives them an edge. While mainland customers go about their day, the Hawaiian-based business owners can sleep, knowing that orders are being captured by e-mail. East Coast orders placed by midafternoon reach Kauai Exotix at the beginning of its workday in plenty of time for next-day delivery to the mainland.

• • •

Remember, too, that anything that can be accomplished online instead of by phone is more cost effective. (See Strategy 3.) A five-minute call to order a $50 item, at a cost of $1 per minute, means that the call is a significant percentage of the cost structure, and a five-minute inquiry—with no purchase attached—creates a financial loss in addition to time lost by personnel who might have been making a sale to someone else. This contrasts with use of the Internet, where—to the extent that communications cost exists—they are trivial, and consumers bear the cost of company contact by paying their access service.

Our idea of what constitutes "expected customer service" tends to evolve with technology. Today, I find myself surprised (and dissatisfied) when companies don't operate twenty-four-hour 800 numbers. The Internet has now led to a new definition of what customers have come to expect: In the emerging era, businesses are almost required to provide twenty-four-hour Internet communications, so that the consumer can shop from home whenever he or she wants to. Sites that prosper will be more than order-taking vehicles; they will provide a creative, educational experience that builds knowledge about their products and services and engenders sales as well as ongoing customer loyalty. (You'll read about how to create this type of selling environment as you continue Part II of the book.)

THE INTERNET SPARKS QUALITY LEADS

Whether your business specializes in water purifiers or ostrich meat, the Web offers you the opportunity to find people who are looking for what your company sells.

In Part I, there were many examples of how the Internet eliminated geographic barriers that previously protected many local businesses and of how new competitive threats have resulted. Now it's important to illustrate how the Internet's geographic breakdown actually presents new opportunities, because suddenly local companies can sell beyond what previously were their geographic limitations. For companies that move quickly to occupy certain niches and for those that sell specialized products for both businesses and consumers, the Internet's global audience creates an opportunity for expanding sales.

Dan Harrison no longer has any interest in reopening the pool supply store he closed in 1991. Instead, he's found that online sales of aftermarket products—chemicals, pool covers, pump replacement parts, inflatable dolphins, etc.—have been a perfect substitute for store ownership. Harrison found that the Web enhanced the direct-sale part of his business by a tremendous amount since he first went online in 1994.

In the business-to-business category, Pall Corporation, a manufacturer of fluid filtrations and purification technologies with 8,000 employees in twenty countries, initially launched its Web site to offer information to its customers, who are generally research and development professionals looking for solutions to their own specific manufacturing problems. What the company found after only the first two months was that genuine sales leads started coming in right away, growing from just a few a week to more than twenty-five after only a few months. Leads generated from the Web site have resulted in customer purchases ranging from $20,000 to $100,000 as the Web has helped Pall overcome some of the obstacles posed by time-zone and language differences.

At one time, retailers that wanted to grow their businesses expanded through adding additional brick-and-mortar outlets. Now, it may be smarter for companies to "go national" by focusing their resources on an expanded presence on the Web.

One example of such an expansion strategy is Holt Educational Outlet, a seller of discount name-brand educational toys, that is using the Web to expand from a local superstore into a nationwide retailer. Revamping the Waltham, Massachusetts, company so that everything that could be automated is now automated, Holt decided to expand through Internet sales as opposed to opening new stores. Thus far, the results are encouraging: Although the company opened its Web presence only a month before the December 1997 holidays, 7 percent of the company's November toy sales came over the Internet. Since then Holt has witnessed a 250 percent weekly increase in e-commerce dollar volumes, and as this book goes to press, the company predicts that Holt will achieve over half its sales over the Web by the 1998 Christmas season. One key element in Holt's success: The company is able to pass on savings to customers because it costs Holt less to do business online.

For businesses selling the unusual, the Internet offers an unprecedented opportunity to reach customers who would otherwise be difficult to find. Steve Warrington, an ostrich broker who sells everything from ostrich meat and feathers to books on ostrich-farm management, runs his business from his Elmwood Park, Illinois, apartment and sells via the Web, reaching 13,000 interested prospects with information on what he's got each week.

Or what about caskets-to-go? A 1994 Federal Trade Commission ruling stipulated, among other things, that funeral home operators must permit customers to substitute their own caskets (instead of buying directly from the funeral home) without a surcharge. While people may feel a bit queasy about stopping in at a casket storefront, an ideal solution was selling online. Catskill Casket Company, which formerly sold only to funeral parlors, now sells directly to consumers after having established both an online and a brick-and-mortar storefront presence. Catskill did not reach this point easily. When the company first started selling direct, funeral directors, who were upset about being cut out as middlemen, boycotted the business and filled Catskill's driveway with returned caskets. Business has come back, however, vindicating owner Joe White's direct-sales approach. In Strategy 10, there is a discussion about sales channel conflicts where direct sales interfere with retailers.

The lessons here are simple. For smart companies, the Internet creates great opportunity, and there are two primary keys to success: a distinctive product and an understanding that it is essential to capitalize on these opportunities quickly. (The value of speed to market is discussed further in Strategy 2.)

THE MODERN TRADING POST

At its core, the Web creates a variety of competitive dynamics because it improves buyers' access to information. One result is that companies are finding that the Internet is exploding as a source for finding alternative suppliers, with potentially lower prices and faster delivery. It is rapidly becoming the world's largest "trading post."

Several companies are working to make it easier for small businesses to compete by giving smaller companies access to the arena of electronic data interchange (EDI) through the Internet. Until recently, EDI systems for business-to-business commerce were prohibitively expensive, and typically only Fortune 1,000 companies could afford such systems. Although EDI makes ordering, billing, and paying for goods and services via private electronic networks fast and dependable, it's expensive to set up and maintain, putting it out of reach for many smaller trading partners.

One example is GE TradeWeb, which is a Web-based service that lets Internet users access a secure Web server to retrieve EDI documents that have been translated into HTML (the universal language of the Internet) or to send HTML documents to be translated into EDI. The only equipment a small business needs to begin trading via GE TradeWeb is an Internet connection and a PC. Users can trade with corporations with which they have already established business relationships, or they can trade with those they see listed on the GE TradeWeb site (including various divisions of GE), customers who formerly would have been out of their reach. As a result, smaller businesses, which could never have approached a giant company, have become supply "regulars." What has been fundamentally transforming about this development is that for the first time small businesses now have a shot at a piece of the action.

GE is a major client of its own new GE TradeWeb service. By opening its bid process to small business, GE lowers its acquisition costs by increasing competition among suppliers while shortening the bid process. In fact, GE Lighting found that this system reduced procurement time by 50 percent and costs by 30 percent.

Once again the lesson here is clear and unambiguous: In the emerging hypercompetitive era, both large and small companies will succeed by offering competitive prices through the lowest-cost ordering process that will include a comparative or "best bid" element. Companies should now be participating on both sides of this spectrum: by using the Web as a trading post to look for new lower-cost suppliers as well as new customers. Companies that wait any longer to adopt this type of approach are, in all likelihood, going to lose market share and valuable sales.

THE BATTLE PLAN

Business owners of companies both large and small can achieve rich improvements in their operations if they start to ask themselves regularly, "I have just been handed a powerful new tool. It essentially lets me costlessly communicate with anyone on the planet. How can I best use it to my advantage?"

For me, this idea is such a powerful and almost overwhelming notion that I believe businesses can do it justice only by stepping back to first principles. To focus, business owners must first ask themselves two questions:

1. As a business owner, what am I trying to achieve?
2. Are there particular "fantasies" I have always had that would make my business work better?

If you spend some time with these questions, and marry the answers to the diverse communications capabilities of the World Wide Web, you will inevitably create some powerful and highly beneficial new initiatives.

Speed Is Everything

I n the emerging economy, the importance of speed takes on a broader meaning than simply being first to the marketplace.

In the past, products were launched and improved in fixed cycles. Every aspect involved in creating and distributing a product was allotted a certain amount of time, and time was also allotted for successfully communicating product information to prospective buyers.

Today the Internet has turned the "fixed cycle" process upside down. Using a Web site, it's possible to distribute detailed information about a product, or even the product itself (such as the latest release of the Netscape browser), to millions of people in one day. For the first time, it is also possible to receive instant feedback from customers on specific aspects of products or services. Companies are now able to watch the activity at their Web sites and glean far more information than ever before about how a product is being received and what kinds of improvements might be valuable.

Moreover, a Web site makes it possible to test out new ideas instantly. I can decide today to change some attribute of a service I might be offering or to alter my marketing campaign. These changes might be permanent or an experimental effort where I'm testing the response of customers and prospects. This type of real-customer real-time experimentation was simply not possible before the Web. Previously, I would have had to spend time and money on focus groups to get this type of feedback.

Why is this significant? Because it is an indication of how companies that succeed now approach the market. We have entered an era of almost-constant experimentation and refinement. In fact, it can be hard to tell the difference between the many small steps involved in a prelaunch trial and the actual launch of a new product.

In an era where there is a constant initiative to enhance, to improve, to stay one step ahead, companies that have mastered the ability to act quickly will have a distinct advantage. They will learn more because they will find ways to experiment more. In addition, they will find low-cost ways to quickly enter markets, assess their relative competitive strength, and in some cases be smart enough to exit just as quickly. Fast-moving firms will leave stodgy, slow, bureaucratic entities in the dust.

To succeed, a company must start with speed as an explicit goal, which is as important as the other central goals that define a company's activities. Speed must be integrated into all aspects of a company: its decision-making process, its training, its attitude toward risk and failure, as well as its management style.

In the competitive environment defined by the Internet, speed is vital ammunition companies must have in their arsenal. And since the Net amplifies and intensifies all aspects of competition, this means that the speed of yesterday is barely fast enough for today. In this chapter, you'll see how speed might be put to work for your company.

FIRST TO MARKET CREATES A STANDARD

Netscape is, perhaps, the most illustrative story of the high value a company placed on speed to market. Because the company experienced some delays with its earliest product, Netscape executive

Marc Andreessen did not want to have this happen again when they were about to introduce their flagship online browser, the Navigator, so he threw down the challenge to his employees: Perfect Netscape Navigator within a month. If they did so, Andreessen promised to appear at the office in Spandex on Rollerblades and eating health food instead of his usual Pepperidge Farm cookies.

A team of programmers took the challenge and began working around the clock, not even leaving for a decent meal or a good night's sleep. One month later, on December 15, 1994, Netscape shipped the Navigator browser, and true to his word, Andreessen teetered in on Rollerblades, sporting black Spandex, and settled down to a big bowl of tofu. This fast launch of Netscape's browser gave it an early dominance in this market.

Any reader who is still not convinced need only look down at his or her keyboard to realize that the dominance achieved by "speed to market" actually existed long before the Internet: The standard typewriter keyboard is a cruelly inferior design that was created in the nineteenth century when the intent was actually to slow down typists because the earliest manual typewriters tended to jam. QWERTY solved this by placing frequently used letter pairs far apart. Though improvements to the keyboard have been created, they have yet to ever make a market breakthrough simply because QWERTY set the standard for the market, simply because it was first.

As evidenced by these examples, the first company to reach the marketplace with a product creates the standard that others must match. While the company may face the difficult job of building a business or service where none existed before, the first out has an unrivaled opportunity to build its brand.

THE FIRST OUT BUILDS LOYALTY
AND DISTRIBUTION OPPORTUNITIES

Any businessperson knows that it's far harder to switch a satisfied customer from an existing product in a category than it is to have a customer try an entirely new product. A satisfied customer needs to be convinced that there is a compelling reason to make a change.

The Yahoo! Net Directory was the first search engine to arrive on the Web. Because a system for Net navigation was so badly needed, it established a following quickly, building a brand with little or no competition. It has retained its preeminence despite the entry of a slew of new competitors (Alta Vista, Lycos, Hotbot, Northern Lights, Infoseek, Excite!), many of which have been heavily promoted. According to Media Matrix, Yahoo!'s market share is essentially double the next search engine's Web site.

What's more, switching loyalties almost always involves a cost in terms of either time or dollars for the customer. One reason Quicken's user base had remained so loyal is that it would take a user enormous time to reenter all of his or her personal information into another piece of software. And the looming threats described at the beginning of this book were, in part, so dangerous for Quicken because they avoided the need for data entry.

The value of being first also gives entrants an opportunity to lock up key distribution outlets. Provident American has established HealthAxis, an online subsidiary. The company has struck deals with AOL and Lycos to be the exclusive direct marketer of medical insurance to their customers, giving it the advantage of being the "first" and, since the deal is an exclusive, of being the "only." The intent of HealthAxis is to leverage cost-efficient electronic distribution channels to be the lowest-cost provider of health care insurance to the individual and small-group market. (The fact that HealthAxis can achieve a low price point reflects, in part, the fact that the claims process operates online.) Like other services, HealthAxis will also benefit from customer inertia. After all, once someone has solved an insurance need, he or she isn't likely to want to switch anytime soon.

Similarly, "first" and "only" tend to go hand and hand in the credit card business on the Web. As a customer convenience, most online retailers put on file the credit card a customer uses with his or her first purchase. As a consequence, being the card of choice when a visitor first enters the site can guarantee a credit card company a great deal of automatic business. In fact, American Express conducted a major online promotion to garner just this benefit.

This kind of loyalty is, in part, based on building a tight relationship with a customer. And if that satisfied customer would have to suffer a "cost" in terms of time or money to switch loyalties away from your product, then he or she is likely to remain yours. (In Strategy 5, you'll learn many ways to strengthen your customers' ties to you, thereby making the option of switching less attractive.)

At least two points are important here: First, companies in the brick-and-mortar world may have distinct advantages if they figure out how to enter the online world with "no baggage." As a separate subsidiary, HealthAxis will benefit from its parent company's expertise but have the freedom to compete according to the rules necessary for winning online. Second, the value of locking up exclusive distribution avenues is high. This should be part of any new product's online entry strategy.

BUILDING A BASE

Pitney Bowes clearly acted on a belief that it should "seize the day" and build a loyal base of customers before competition arrived. For nearly eighty years, the company has dominated the postage-meter business, principally among large companies. Now, E-Stamp and others have developed nascent technologies that let them sell and distribute postage for home offices and businesses over the Internet with indicia that are printed via the computer.

Pitney Bowes recognized that the technology was here and competition was coming for the 50 million businesses with fewer than twenty employees. Opting to stave off competition from others developing Internet products, Pitney Bowes decided to get there first and work to dominate the market by creating a non-Web product, the Personal Post Office, a new style of postage meter that provides the same user benefits for home office workers and small businesses. The attraction to the product is that users can "refill" the postage meter over the phone lines. The company has been particularly aggressive at marketing Personal Post Office, and Pitney Bowes has priced the product at a reasonable $19.75 per month (or $24.75 per month with a separate scale) plus a sixty-day free trial and $25 in free postage. Using the product also provides some significant savings: There is no wasted extra postage, since you are able to weigh your mailing and set the meter for the exact amount needed.

While eventual competitors like E-Stamp and others are identifiable but still in development, Pitney Bowes has the market to itself. As a result, the company is working aggressively to build share before competitors come in. If customers already have the Pitney Bowes Personal Post Office meter in their homes or offices and are satisfied, it will be quite difficult for new competitors—such as E-Stamp—who lack a brand name or obvious product advantage to succeed in the business.

Personally, I am a Personal Post Office customer, and I have found it very useful. At this point, the likelihood of my switching systems is minimal.

When companies are competing hard against each other, it's generally good for customers. The threat of competition often leads to lower prices and constantly improving services. It's impossible to know, but would Pitney Bowes have promoted its product so successfully or priced its Personal Post Office so affordably if clear competitors were not on the horizon?

In the evolving hypercompetitive climate, if a company has the ability to build a large customer base before the arrival of direct competitors, it should do so with a full-out assault. Otherwise, a valuable opportunity will have been lost.

FIRST OUT SETS THE RULES
AND GETS THE FEEDBACK

There are additional advantages to creating a market: Whoever follows the first out must play on the terms established by the initial entrant. To build market share, the new competitor must take a product or service to the next level. When you're number two, "as good" isn't good enough; you have to be better, demonstrating why consumers or businesses should buy from you instead of the established industry leader.

If the first out has established a low price point for the product, number two must accept that range or justify a higher price. Similarly, the first entrant defines a quality of service that the new entrant must now match—or explain why it is not matching—while charging a substantially lower price or offering other benefits.

CDNow, an online music retailer, knows that competitors are coming at it from all sides. As a consequence, cofounder Jason Olim says they will follow their mantras "Get big fast" (before more competition comes in) and "Speed is God." By maintaining strong customer ties and a huge selection, Olim does not expect this battle to be one involving price; he anticipates that faster and better should keep CDNow ahead of the pack.

In addition, the sooner a product is out, the greater the opportunity for the manufacturer or service provider to learn about and refine the offering. The company will be able to incorporate what it learns and consistently stay one step ahead of the competition. While the competition is first learning about this market, the early entrant is using its experience to refine its product for its next version.

One person I met with several times while working at Time Warner was Charles Conn, the CEO of CitySearch, one of the leading online city guides. We met to explore potential alliances between our companies. He talked with me about another compelling reason for wanting to move quickly. Conn believes that "long lead times compound any mistakes you are making." As a result, in a "live or die" hyperbattle, if you get to market quickly with a prototype, you can make mistakes and correct them *before* they sink the company.

CONFRONTING THE ORGANIZATIONAL CHALLENGE

The management of barnesandnoble.com believes that continuous, fast development of new products and services is critical to success in a hypercompetitive environment. In fact, the site strives for a major new product development every forty-five to sixty days, with smaller improvements taking place on a continuous basis.

This ongoing dedication to speed creates a serious challenge for any organization: how to continuously innovate in a competitive environment without overworking and burning out employees. Several of the approaches that are employed at Barnes & Noble's online effort include:

First, executives work to create a mind-set that embraces this philosophy throughout the organization—what one executive describes as "an understanding and dedication to rapid product development and cycle times."

Second, specific product releases are assigned to specific teams. In many companies, the top technical people are often sought for every task. As a consequence, the best people become overworked and are the least happy. Executives work to avoid this problem by making assignments carefully and keeping teams focused on their specific tasks.

Third, executives strive to avoid at all costs "the stop and start" phenomenon. Once a team begins a task with a clear understanding of what they need to accomplish, he steers away from the natural impulse to stop the group and move them to something that appears to be a higher priority at the moment. If priorities change every day, nothing is ever accomplished without difficulty.

And fourth, executives work hard to have a clear idea of where they want to be six months from now so that they can provide their organization with appropriate leadership. As discussed in Strategy 9, executives at successful companies have a clear sense of priorities and where they want to go. As a consequence, they are able to provide their organizations with strong direction. With clear priorities, the appropriate people in an organization can focus their energies.

One area this book does not address in depth is the human side of HyperWars. As companies attempt to maintain a rapid stream of innovations, it's easy for "technical people," on whom all of this depends, to burn out. In a hypercompetitive world, there is rarely a sense of completion or a time when the pressure is not high. This was one of the central challenges I confronted in building a large-scale commercial site. What's essential is that company managers recognize the important dimensions of this issue, and have a clear plan for addressing it.

COMPETING FOR NICHE MARKETS

In an age of extreme competition, available market niches open and close rapidly. Fast-moving companies win in this type of environment for two reasons. First, they are swift enough to see a niche and respond instantly. Second, because they grabbed the niche, they are in a strong position to continue the innovative process around this market niche. By getting their feet wet before the water warms up, they learn to swim before the other guy. As a consequence, they swim up the learning curve in any given area at a much faster pace.

LESSONS LEARNED

In practical terms, "speed is everything" means:

1. It's more important to get out first with a good product than it is to wait to try to make the product "perfect."

Of course a product or service must meet standards of quality. However, as windows of opportunity are limited by time, it is generally most important to get out, get going, and start learning. If you are first and you are valued, there is not a higher quality standard against which you will be measured.

While working for Time Warner, I learned about "speed to market" firsthand. We decided to develop a paid subscription personal news feature for Time Warner's Pathfinder site, but we spent over a year designing and building the product so that it could handle huge volume. Our mistake: We knew how we wanted the product to evolve over time with different and compelling features, so we delayed while trying to make everything fit. In the meantime, our potential lead time over our competitors was significantly reduced. The lesson: *Get to market first, then keep modifying the product to stay ahead of the competition.*

In fact, I felt this lesson so strongly that when I formed my own company and developed the idea for my online newsletter *Bruce Judson's Grow Your Profits,* I pushed myself to move from developing the concept to a full-scale launch in only two months. My reward: an open territory for the product and press recommendations that recognized its innovative nature.

2. Since product opportunities have limited time horizons, it may be smarter to acquire companies or people with experience than to try to develop them yourself.

Cisco, the leading provider of networking products, is the classic example here. The company is constantly acquiring businesses so that it can move quickly into new areas and is quite open about this strategy. It makes these acquisitions—now numbering over twenty-five companies—because it believes it does not have the time to develop its own expertise and get to market with integrated products fast enough.

3. Don't enter a market unless you have the capabilities to continuously evolve your product entry. If a product is outside your core business, you may not be able to move quickly enough to keep it competitive.

If I return to my two contrasting experiences, this point is reinforced: At Pathfinder we lacked the technological depth to continuously enhance Pathfinder Personal Edition, a personalized news service, at a rapid enough clip. In contrast, I made sure that I had the resources and a migration path in mind before I launched the first issue of *Bruce Judson's Grow Your Profits*.

4. Form partnerships for speed.

Later in this book, I discuss the important role partnerships and alliances can play in HyperWars. Here, I simply want to point out that it may often be easier to get to market faster by working with a partner than by working alone. Your partner may have specific skills or a knowledge base that your company lacks.

SOMETIMES "FASTER" IS AS GOOD AS "FIRST"

In our evolving, intensely competitive world, it's clear that speed wins. Speed can be a competitive advantage, either in being early to market or in being faster than your competitors in whatever business you are in.

It's worth noting that the Internet is also a means of enabling companies to gain a competitive edge by increasing the speed of their activities and ultimately the speed with which they can provide services or products to customers.

The following are examples of companies that have figured out how to harness this capability to their advantage:

Adaptec

Adaptec, a $1 billion chip maker, used the Internet to reduce product delivery times that were nearly twice as long as those of competitors. Adaptec had taken up to fifteen weeks to deliver chips to some customers, including industry leader Compaq. Competitors, such as Cirrus Logic, were taking only eight weeks. Cirrus was able to move quicker because it operated its own manufacturing facility and could optimize production schedules. In contrast, Adaptec worked with three outside partners for manufacturing.

One solution Adaptec adopted was to create an extranet (an electronic link that a company can open to its suppliers and customers) that allowed all of its partners and suppliers to integrate more closely. By using e-mail to generate and confirm orders and by sending new chip design diagrams over the Internet, the company was able to ramp up more quickly for product changes and new designs. By revamping its own procedures, Adaptec was able to make changes that allowed it to equal the competition's cycle time. "In this business, you're either fast or forgotten," an official of Adaptec was quoted as saying in *Business Week*.

As Adaptec found, speed requires a major commitment within the company. Previously, paperwork at Adaptec took two weeks to get into the system. This two-week period provided a cushion where managers could change their mind about a particular order. Now, final decisions about products need to be made faster so this cushion is eliminated—the staff must make a final decision fast, the first time.

Eli Lilly

In the pharmaceutical business, it can take a decade and hundreds of millions of dollars getting a new drug to market. Yet a key value in getting drugs approved faster has to do with income: Drugs are most profitable when they are sold exclusively under patent protection.

Eli Lilly shepherded a new schizophrenial drug through the approval process in seventeen countries in only eighteen months by using its intranet (an online presence accessible to your employees but closed to outsiders) to link sophisticated databases that provided constant communication and responses to questions. The company shared knowledge on a worldwide basis and was therefore able to leverage what it learned in one country, in others. The benefit? Zyprexa generated over $550 million in revenues in the first fourteen months following its multinational approval in September 1996. Eli Lilly is now using its intranet to support research sharing to speed up new drug discoveries.

IBM Credit Corporation

The IBM Credit Corporation has established an Internet-based service that allows its customers — computer remarketers — to close financing deals in minutes instead of days.

"In our highly competitive business, speed wins," says Chuck Thomas, director of global channels customer financing, IBM Credit. "Our new Web-based tools will dramatically improve the rate at which our business partners can obtain the financial information they need." The Web site adds tremendous value to the channel by providing information that simplifies the quote process. As a result, IBM Credit takes a major step in providing top-quality value to its customers.

Truxigns

Truxigns, a Walnut, California–based sign-design company, once relied heavily on FedEx to send and receive photos and designs that the company places on the sides of fleet trucks and vans. Now the Internet is the medium of choice for delivery, lopping two full days off some jobs by cutting out the need to overnight designs back and forth from clients. "In our business, how fast you can serve somebody has a lot to do with their willingness to use you," says a company representative.

THE BATTLE PLAN

In the emerging era, a key factor in the ongoing success of businesses of all sizes is speed: both the speed of the service provided to consumers or other businesses and the speed with which new or enhanced products and services are brought to the market.

I have found that there are some consistent approaches that lead companies to develop the capacity to act quickly. First, speed must be widely recognized as a valuable goal in and of itself. Today companies ranging from successful start-ups to technology giants actively work to encourage a corporate ethos that values the ability to create and act quickly.

Second, companies with an appropriate focus on speed also have the same focus on their competition. They know how long it takes their competition to fill a customer order or to bring a new product from concept to market. They are vigilant in constantly watching to be certain that their competition has not done something new "to break the sound barrier."

And third, companies with a focus on speed allocate valuable resources—money, time, and talented people—to experimenting with ways to eliminate time from their operations. As they are watching their competitors, they are also working to develop new methods of their own.

Cut Costs and
Increase Efficiency
Using the Web

Today almost no one is making money on the Internet" is a frequently heard comment about the Web. However, this chapter will demonstrate that the comment is wrong. Generating revenues is one aspect of using the Web for business. Another important aspect is using the Internet to reduce costs. The end result? Many, many companies are already profiting from the Internet by using it for cost savings, often making a significant difference in the bottom line.

Creating a cost-competitive organization is an important ingredient for success in this new era. As price competition intensifies, businesses with lower costs will be able to compete the most aggressively while maintaining profit margins.

Consider what's happening in these industries:

In the low-margin grocery industry where saving pennies counts, Harmon Supermarkets, a privately held upscale $250 million chain, expects to see a substantial profit increase due to the savings realized by its new extranet. Before, the process for repricing grocery items was painstakingly slow. A weekly report from the wholesaler would have to be reviewed item-by-item by store personnel, with each store then having to feed its updates into the data feed for checkout scanners and shelf labels, thereby creating a significant expense in staff time and cutting profits because of the lag time in getting new prices posted.

With Harmon's new extranet, price changes, which formerly took a couple of days, can now be implemented instantly, passing on cost reductions to consumers immediately, or preserving profit margin on goods where the supplier has increased the price. The process saves staff time, lures in more customers for better prices, and maximizes the period of time when the "manufacturer's special" can be offered. The extranet can even be specially tailored for each of the chain's stores; the store profile can preset a rules-based system that lets the computer know exactly what profit margin a local Harmon Supermarket wants to preserve on a can of peas or by how many pennies it wants to undercut a competitor on a box of cereal. The result is a competitive edge for Harmon as well as a big increase in profit.

The possibilities for online cost-savings are unquestionably going to expand at an exponential rate over the next few years. Brutal competition and price pressure will require that companies take advantage of every possible type of efficiency available. The faster companies integrate these new techniques, the sooner they'll be able to use them as weapons against competitors, who are likely still at the beginning of the learning curve.

While Strategy 1 focused on using the Internet to improve communication, in this chapter, the focus is on how to save money using the Internet. In other words, if Strategy 1 was the "better," this chapter is the "cheaper."

REDUCE SELLING COSTS

Merchants of all kinds, especially those where sales are primarily solicited via catalog or direct mail, have greeted with excitement the thought that there is an increasingly effective way to deliver information to masses of customers without having to pay printing and mailing costs. Those with foresight are realizing benefits from online activities.

Kodak has greatly reduced its print run of product catalogs in recent years by letting remote sales agents use its Web site as a print-on-demand service. The agents enjoy the added benefit of constantly updated information, while Kodak reaps significant annual savings.

Online catalogs and other publications have multiple virtues: They generate obvious cost savings, they can be updated on a continual basis, and they can include a variety of unique cost-saving features that are expensive or impossible using paper-based alternatives. As buying habits shift among both consumers and business-to-business customers, online catalogs and transactions will become as central to doing business as paper catalogs and business forms are today.

AUTOMATE ONLINE ORDER
(AND APPLICATION) CAPABILITIES

Whether it's a consumer sitting down to place an Eddie Bauer order, a teenager zapping in an order through CDNow for the latest Fiona Apple CD, or someone sending flowers via the Internet, the thought that companies can reduce telephone charges (which are typically $1 per minute for an 800 number), eliminate costly errors in the order process, and reduce staffing costs is very compelling.

One of the most innovative direct-order businesses to pop up on the Web has been iPrint, a company whose chief executive quickly realized that if the company combined the design capabilities of the company's proprietary computer software with the ordering potential of the Web, it could create a design-and-order stationery service and electronically link it to a commercial printer that could then process the order. iPrint allows customers to design high-quality business stationery online, and its automated system represents a real threat to the local printing industry. The company's prices are almost 50 percent lower than those of traditional printers. iPrint's methods and linkage directly with printers reduce costs because fewer people are involved, and its automated online design (What-You-See-Is-What-You-Get) process reduces between 16 and 20 percent of printing errors, thereby eliminating the need for redoing print jobs—a major cost for traditional local printers.

1-800-Batteries is happily moving toward more Web-based sales. The company launched its site with the thought that consumers who are fully equipped electronically need easy access to batteries for their laptops, their cell phones, their beepers, and a multitude of other electronic accessories. What the company has experienced is a decreased cost-of-doing-business, which it is now sharing with consumers. While phone orders cost about $14 each to process, Web orders cost less than $7 and have fewer errors, so the company now offers a $2.50 discount for online orders as a way to entice buyers to do business on the Internet.

And, of course, the lower-cost Internet-based brokerage services have led to a whole new pricing structure for the industry. Here lower costs translated into an entire new product line that is "taking over the world."

And don't worry. It won't be long until you're invoicing and accepting payments online, too. Estimated savings range from 20¢ to 50¢ per single billing over the paper, postage, and processing that now goes into the nearly 19 billion invoices mailed to American consumers each year. Such savings are expected to spur growth toward 2 billion bills transacted online within four years.

It may take a while for consumers to become accustomed to ordering and paying bills online. Indeed, consumer habits change slowly. But this shift will inevitably occur. The potential savings for businesses are so significant that we can anticipate financial incentives for consumers (such as significantly lower prices online) or other benefits that will serve to accelerate this transition.

ADD INTELLIGENT SOFTWARE

While all this is quite remarkable, the skeptics are still going to say, "My business is too specialized. Customers won't know what to order if they don't talk to someone." Of course, at times, this is all too true; however, some software is so sophisticated and easy to use that it actually outperforms the personal touch, simply because it eliminates a costly factor every business must deal with—human error. Consider this:

What could be more complex and personal than ordering a computer networking system that will successfully integrate with a company's own system? That's what Cisco sells—and it's successfully making a lot of those sales online, thanks to the development of "truly intelligent" software that incorporates rules related to custom product configurations.

Cisco's Web site allows customers to configure products with complex features, and its "intelligent configurator" software rejects orders where specific components won't work with other parts. As noted earlier, inaccuracies on orders processed through the Internet have actually dropped from more than 25 percent before this automated system existed to less than one-tenth of a percent now. Annual savings from Cisco's site, which reflect the elimination of order errors as well as other administrative efficiencies, are estimated at over $250 million a year, and the absence of errors has also allowed Cisco to improve delivery time by three days. For Cisco, the Internet has meant faster service, quicker production cycles, and savings.

ESTABLISH MAXIMUM VALUE
FROM AN INTRANET

An intranet for your business can keep staff fully up-to-date with everything from personnel benefits to the latest data they need on sales. Consider how an intranet has changed Greyhound:

No one would automatically link "bus service" and "cutting-edge technology," but Greyhound, a company that must maintain control over 2,000 buses and more than 4,500 drivers, has embraced the future and found that in doing so, it has established a strong base of management and saved money at the same time. Through Greyhound's central dispatch office in Dallas, its BOSS (Bus Operational Support System) essentially orchestrates buses and drivers nationwide.

In the past, Greyhound's dispatching operations were controlled from a number of regional centers, and there were major inefficiencies in coordinating a nationwide system. When Greyhound overhauled the system, regional offices were eliminated, and now drivers receive their daily orders via the BOSS intranet. When drivers arrive at their terminals, they log into BOSS to determine their assignments, and just prior to departure, drivers input final passenger counts, which are then immediately transmitted to Dallas. This alerts supervisors to buses that are filling rapidly and subject to overflow. Then buses can be redirected as needed. A spokesperson notes that now about eighty to ninety dispatchers perform the functions that formerly were done by several hundred. Drivers' pay is going up because the fleet is being used more efficiently and drivers are getting more hours. One day drivers will be able to log on from home.

Denver-based U.S. West, a regional Bell operating company, is devoting millions of dollars to support a forty-person SWAT team, the Global Village Labs, whose mandate is to save tens of millions of dollars for the company. The key to the savings realized lies in the creation of its intranet (the usage of which is detailed in the next section). Thus far, the savings have been nothing short of impressive. According to the company, the work carried out by the Global Village saves several million dollars a year.

While drivers likely aren't your problem and you have yet to create a SWAT team, an intranet can still likely save your company money. I've found that each day I seem to find another compelling way to use an intranet.

By presenting you with the following examples, I hope it will spark your thinking as to what you can do within your own company. I've found great savings to be gained primarily in five areas:

1. Training and sales support
2. Printing costs
3. Up-to-the-minute information
4. Personnel issues
5. Knowledge management

Training and Sales Support Among the ways U.S. West is reaping savings is through its new employee training system. Its Global Village SWAT team has created what they call "just-in-time" training. Instead of pulling someone out of the workforce and putting him or her in a conference room with an instructor, they've "Webized" their training so that it can be done over the Web for fifteen minutes between shifts or during downtime on a slow day. The Web lets people connect to lots of self-help, self-learning opportunities. Training is now more efficient and less expensive and also more responsive to employee need. (In the days of long-term employment, it was standard to consider employees "in training" for two or three years, since they tended to stay for a lifetime. Now it's vital to train staff quickly because they are more likely to move on.)

Knoll Pharmaceutical Company in Mount Olive, New Jersey, launched an intranet designed to save the company $1 million a day. Prior to establishing this internal network, Knoll needed to bring its staff of 600 field salespeople to at least three national meetings per year and more frequently if a new product was to be launched. In addition, many smaller local meetings were necessary. Knoll's sales training director says the intranet was designed to eliminate the need for as many sales meetings (reducing travel costs and lost selling time), keep the sales force more fully up-to-date, and reduce paper costs. (Company officials calculated this high potential value for its intranet by looking at savings in travel and paper costs as well as revenues that were lost when the sales force was called in for meetings.)

Printing Costs Merrill Lynch is using its intranet to reduce the cost of printing and distributing the truckloads of research reports and manuals that its 27,000 employees must follow.

Eli Lilly is using its intercompany electronic linkage to unite more than 25,000 employees in thirty countries; its intranet contains more than 12,000 pages of information ranging from employee benefits communication to sales and marketing databases. In addition, it also provides consistency of information and pricing throughout the company's offices.

After two years of crafting and implementation, AMP, the world's largest manufacturer of electric connectors, launched its online catalog to service its 90,000 customers in fifty countries. The Harrisburg, Pennsylvania–based company saved 40 percent on an annual budget of $15 million to publish and distribute paper catalogs worldwide. The site has also helped the company significantly reduce its annual fax bill.

Up-to-the-Minute Information Konica Business Machines USA is enjoying dramatic benefits from its intranet. A company for whom a flashing "document jam" light on customers' copy machines means trouble, Konica (where the staff turnover rate is every eighteen to twenty-four months) had difficulty keeping the technicians fully up-to-date with technical developments—it was both costly and time-consuming to pull workers out of the field. What's more, most of the technicians know the machines well enough that what they generally needed was to be pointed in the right direction, not to be put into a classroom to be lectured as to how to get there.

Konica began using its Web site to make available a knowledge base of 48,000 possible solutions to 5,000 dealer technicians in

North, South, and Central America. (Technicians can access the information via computer, or call the help line, whose staff will use the database to help solve the problem.) A user can type in a problem, and the system will help refine the question and recommended solutions. According to the company, about 82 percent of the incoming calls can be answered using the database. Konica envisions the migratory path for this capability: In the future, Konica will provide some technicians with laptops so that they can access the Web from the customers' sites; the company also intends to eventually open Web queries to customers.

Personnel Issues Human resources departments reaped the earliest benefits from corporate intranets. Posting the employee handbook, the company newsletter, and job openings was the first and most logical application of the technology. A few pioneers began using their own self-service human resources applications for the intranets, and now there are excellent software programs that help any company who wants to develop self-service applications, self-administered 401(k) programs, and other human resources applications.

Santa Clara, California–based Applied Materials, Inc., a manufacturer of semiconductor equipment with 13,000 employees around the world, found that its paper-based benefits-enrollment procedure for newly hired employees was a long, involved process that resulted in errors, incomplete forms, and illegible handwriting. Since late July of 1998, it has enrolled 1,000 new employees using an intranet-based set of self-service applications. A new employee can enroll him- or herself in the benefits program in about five minutes, and the personnel department reports that errors are low and there has been no need for staff to rework any of the forms. Because not all of its employees (factory workers) have access to desktop computers, Applied Materials is constructing intranet-based kiosks for only $3,500 each, eventually making the intranet easily available to all.

Employees and corporations alike are delighted with the new possibilities offered by filing expense reports via the intranet. To employees, the easier it is to input business expenses, the faster they can receive their reimbursements. For organizations, the automating of expense reporting promises to reduce the cost of the process exponentially. NationsBank expects to save half of the $85 it costs to process an expense report, and the company handles more than 30,000 expense reports a year. For many companies, this system has simplified what was before an unwieldy process.

Knowledge Management Holland & Hart, a full-service law firm with over 200 attorneys located throughout the Rocky Mountain region, wanted a way to provide its staff with fast, easy access to relevant information. The data had already been converted to an electronic format via word processing; the question was how to share all that research with all of its staff.

Today, using an intranet created by Verity, H&H attorneys and research staff now have access to tools that enable them to organize, search, and manage all the firm's information resources. With over 600,000 documents online, an H&H attorney can find an expert witness from a case that was closed several years ago, a process that would have taken days or weeks previously, since the search would have to be done by hand. H&H also plans to use the search capabilities of its system to reduce costs for other litigation needs. An attorney can access the H&H intranet to search for previous courtroom re-creations, graphics, videos, and animations that might be relevant. This ability to tap into resources on a collaborative basis should bolster a client's case while saving money at the same time.

Historically, GE Capital's twenty-seven different business units employing 52,000 people existed as autonomous entities, but with the creation of its intranet, the company sought to drastically alter the corporate culture. The challenge was to create a system that enabled employees at all levels and in every department to share their knowledge and skills so that the entire company could profit. Incentives seemed the answer. To overcome the long tradition of internal competition, the company established a practice that informally based promotions and raises on how well employees shared their knowledge online. The results have been worth the push. It is estimated that having access to other business units' projects has reduced the workload of hundreds of employees by 50 percent because they no longer have to start at "square one" each time they create something.

In many ways, we are still in the "first generation" of intranets. Therefore, it is inevitable that new ideas and developments will provide even greater benefits than those described here. Anyone in business should investigate, and create, an intranet if they have not already done so. Without putting the right tools in place, it's hard to imagine that firms will successfully battle for market share against fast-moving companies that effectively harness this new technology. Moreover, the development of an effective intranet creates an ongoing learning process. As companies deploy new intranets, employees almost always discover additional effective uses for the intranets that *were not imagined* at the time the intranet was initially developed. Only by taking the important first steps of establishing this vital business tool can companies benefit from this type of learning.

ESTABLISH AN EXTRANET TO LINK YOUR COMPANY AND ITS DATABASES TO MEMBERS OF YOUR EXTENDED FAMILY, PARTICULARLY YOUR SUPPLIERS

Extranets, Internet links with a select list of customers, are opening up cost-effective avenues of deepening business relationships.

Sharing the wealth has been the key to Chrysler's mega-savings. Its supplier cost-reduction efforts have led to a stunning billion-dollar savings a year on operating costs. In 1997 Chrysler's total profits were $2.8 billion. To put the value of the extranet in perspective, I calculate the cost savings achieved by Chrysler through this well-managed innovation accounted for over 30 percent of the company's total profits.

Under the program, suppliers submit ideas online. If Chrysler accepts and uses an idea, the supplier not only gets the business but also receives a cut of whatever cost savings is achieved. So far, the ideas have ranged from ways to redesign tail lamp assembly (savings, $174,000) to ways to redesign car airbags (savings, $3,047,098).

Large companies may also find it beneficial to launch extranets that link them to their dealers. Toshiba's fax manufacturing subsidiary recently launched an extranet designed both to save Toshiba costs related to its order system and to provide dealers with a reliable means of overnight ordering, thereby reducing their inventory needs and costs. By setting up a Web-based parts-and-supplies purchasing network for its 350 dealers, Toshiba expects to cut in half its existing online selling expenses (from $1.3 million to $600,000), saving more than it did with a proprietary system.

The system provides far superior service to dealers, because the dealer can receive a confirmation immediately, get an order number, and check order status any time of day. Links to FedEx tracking show shipment status. (Toshiba provides the dealers with a Netscape browser and local Internet access from MCI.) Because of prompt delivery, dealers don't have to stock as much in inventory because they know they can get everything right away.

Earlier, we discussed how companies that start down the appropriate Web "path" tend to accelerate their activities as they realize ever more benefits. Extranets reflect this phenomenon. Companies reaped so much benefit from their intranets that they opened their systems to their extended families of suppliers and customers. If your company has not yet investigated this, the time to do so is now.

INCREASE YOUR OPPORTUNITY
TO FIND FIRST-RATE SUPPLIERS

Many companies are having great success with establishing Internet-based procurement systems at their Web sites to encourage "new blood" among their suppliers. GE, through GE TradeWeb, conducts large portions of its supplier bidding online, lowering acquisition costs and increasing competition among suppliers, while shortening the bid process. GE expects to save $500 million by the year 2000 through online bidding and the accompanying increase in potential suppliers.

And Sony is working to set up an online system to reduce parts-procurement costs by sharing quality-control information with over fifty domestic and overseas parts-procurement offices. The system will cover 1 million parts procured from 10,000 domestic and overseas firms. Because the system will offer instant access to information such as the defect rates of parts, it is expected to lead to savings on the order of hundreds of millions of yen.

This form of procurement will not be appropriate for every company. However, experience to date suggests that many companies that did not anticipate significant savings found that when they put their toes in the water of online procurement efforts, new suppliers emerged along with significant savings.

MANAGE PURCHASING VIA THE WEB

Forrester Research, which studied fifty large companies that have already moved at least $100,000 of purchasing to the Web, found that these companies are placing an average of 12 percent of all expenses related to maintenance, repair, and operations orders over the Internet. However, Forrester projects that by the year 2000 these same companies will be placing 55 percent of orders— $32.4 billion—over the Internet. In the process, they are reducing transaction costs, minimizing the time from order to delivery, and freeing up purchasers from mundane transaction-processing work. And figures from market-research firm Killen & Associates pegs the cost of a single paper-based and -processed purchase order at about $144; an electronic purchase order done via a company intranet, and using a preapproved catalog of company suppliers, costs only $5 to $40.

Today companies can still maintain control and the resulting bargaining power, but company intranets permit each department to buy supplies from preapproved suppliers with whom the central purchasing department has worked out the best price. Companies essentially have their own online catalogs from suppliers that each separate department can access—but the company gets the benefit of prenegotiated prices or volume sales.

For employees, an online procurement system works like this: The maintenance staff needs more mops or a manager needs a laser pointer for a presentation. The staff member logs on to appropriate software that launches an Internet browser to log on to the procurement site on the company intranet. A requisition form can be requested, and then the employee navigates through the employer's custom electronic catalog of all the goods and services that employees are authorized to purchase. After clicking on the appropriate selection, the form is automatically routed to the staff member's supervisor for approval when appropriate, and then it's forwarded to the wholesaler or manufacturer of the item. The appropriate department is charged for the cost, and at any time, the employee can check on the status of the order.

Visa International is pleased with the online procurement process: The company noted that recently a procurement order went through the process in thirty-three minutes versus the three to five days it would have taken previously; and Visa is saving 50 to 80 percent on handling of each of its purchases.

Systems that allow this type of decentralized purchasing with the corporate clout of prenegotiated prices are not cheap, and costs can vary significantly based on the complexity involved. However, the annual savings for a company with $1 billion in sales can reach $5 to $10 million, which can be 1 percent of total revenues. If a company operates on a profit margin of 8 percent of total revenues, this means that this one initiative could by itself boost the bottom line by 12 percent. A company with $1 billion in sales and $80 million in profits that successfully implemented a procurement system of this type would now realize $90 million in profits. This large and immediate payback clearly makes these efforts worthwhile.

THINK GLOBAL SALES, WITHOUT A LANGUAGE HASSLE

One benefit of the Internet is its ability to facilitate sales on a worldwide basis. For example, Marine Power Europe, a boat engine manufacturer, deployed a system that functions in eight different languages and lets thousands of independent agents search inventory and check status of orders in their native languages, thereby reducing customer service costs for this international company. Prior to the extranet, the European agents typically retrieved information by telephone or by fax.

EXPLORE NEW AVENUES OF PRODUCT DELIVERY

Business history strongly indicates that at the end of the day the lowest-cost form of distribution wins. Either existing businesses embrace new, lower-cost methods, or upstarts use these methods to underprice and outcompete an industry's established players. It is, therefore, entirely reasonable to expect that the Internet will increasingly become a mechanism for distributing new software, software updates, and entertainment products such as music directly into the customers' computers.

This phenomenon is still in its early stages, but its growth is inevitable. As high-speed access becomes available to consumers, this form of distribution is likely to explode. As a consequence, companies involved in the creation and distribution of software or music need to start positioning themselves—by establishing online plans and beachheads—for this forthcoming, sweeping transition.

All types of software are already being delivered over the Internet, and within the next few years, it will be common for customers to pop their recordable CD into their computer, download the album or single they want, and pay for it in cyberspace.

While many companies are appropriately wrestling with how to bring in new and growing revenue via the Internet, there is no doubt that used properly, a Web site can increase efficiency and reduce enough costs that corporate accounting departments are going to get greedy.

THE BATTLE PLAN

For businesses large and small, the potential savings from deploying many of the new cost-saving opportunities merit a systematic approach.

I would suggest that every organization assign an individual or department with the responsibility of keeping the company at the cutting edge in cost savings. This involves three responsibilities:

1. Devoting energy and time to staying current on developments in this arena
2. Assessing whether implementing these new developments makes sense for the organization
3. Providing an analysis of the costs and benefits that is as precise as possible when recommending new initiatives

Progress should be reviewed by company leaders on a monthly basis.

A Great Product Today Requires Something More

When I speak to large business audiences about competing in the emerging online era, I sometimes note, "It's essential to have a product that's valuable for your customer; unfortunately, a great product is not a certain route to success." Success today requires a commitment to providing a valuable product *and* using smart strategies to engage the customer.

At one time, it might have been enough to develop a great product and bring it to market with a well-orchestrated launch. Now, competition requires that to succeeed a company must also master the variety of related skills discussed here—ranging from reacting swiftly (speed is everything) to personalization (an ongoing initiative to show how your product meets the needs of specific consumers).

The Internet expands our concept of what is a product or service. The Web may make it possible for you to improve your product or service in a way that wasn't possible before, but it can also play a crucial role in determining how a customer experiences your product. The Web may be the first place a prospect interacts with a company and the first place an existing customer visits for product support when something goes wrong. In effect, the Web should be thought of as an extension of the product—with all of the attendant attention to quality and customer service that this implies.

To compete in this fast-changing environment created by the Internet, it's important to keep in mind that success depends on the product you are providing your customers as well as on how you make it available to your customers. This chapter expands on these ideas.

THE PRODUCT EXPERIENCE IS THE PRODUCT

A number of management thinkers have discussed the idea that for the customer today the product is more than the product: It is the total experience the customer has in interacting with the company involved. This runs the gamut from how the company provides product information before a purchase to the postsales customer service the company provides. In general, it is this total experience that will determine a customer's loyalty.

Because the Internet creates new ways for customers to learn about, shop for, and buy products, it makes the total customer experience that much more important. Customers can buy life insurance or lawn furniture almost anywhere—the Web site that gets a customer's business is likely to be the one that offers excellent service, fast delivery, or best price, and sometimes all three.

Within this context, I have isolated four questions business owners can ask themselves in evaluating how the Internet may affect their ability to deliver a competitive product:

1. The tsunami effect: Does the Internet threaten to undermine your core business?

For each industry, the impact of the Internet will be felt in different ways and at different times. For all but a few industries, the Internet is like a tsunami: a giant, unstoppable wave that will sweep over the horizon with tremendous force, forever changing the landscape of everything in its path. The changes it is bringing about are so fundamental that you may discover that it has affected your core business, or even your entire industry, as it recedes.

"Can this happen to my company?" is a question every business-person should ask. "When the tsunami strikes, will I be on high ground, or am I at the shoreline, about to be overwhelmed?"

Think of what's happening in e-commerce sales: Items as varied as housewares, groceries, electronics, industrial components, and garden supplies are actively being sold on the Web. With this in mind, you realize that online commerce can affect virtually any industry. To help you evaluate how close change is coming to you, consider the core question posed in Part I: "Can a product be sold better, faster, easier, or cheaper on the Web?"

If the answer is yes for a product in your company, then you'd better start making plans for evolving the product fast.

There is a human tendency to say, "But my customers are loyal," or "Distribution channel conflicts will prevent that," or even "It won't happen before I retire." Don't be deceived. No one anticipated the extraordinarily rapid transformation the Internet has caused in a wide variety of industries, and the difficulties it has created for brick-and-mortar businesses in these industries. Today you may successfully operate a brick-and-mortar storefront, but before tomorrow you'd better know what you want to accomplish online.

The key to surviving and profiting through this change is to act quickly and directly:

- Acknowledge that dramatic change in your industry may happen at any time and that "business as usual" is no longer good enough.
- To the maximum extent possible, get out in front of the changes that are occurring. It's easier to evolve a business if it happens on your terms (a proactive approach) as opposed to those dictated by new competitors (a reactive approach).

- You can't be afraid to cannibalize your business by launching new and better products—even while the existing ones are still healthy. For example, getting in front of the wave may mean sacrificing a higher-margin product today in order to establish a leadership position for the evolving world of tomorrow.

At this writing, Egghead has beaten a hasty retreat from retail outlets in favor of an online emporium. This reflects a need to rethink the company's entire business. The software business is intensely dynamic and competitive, and the company's shift is no guarantee of success. New sweeping changes (such as the ability to rent software online) are likely to further redefine the competitive environment, and there is no guarantee that Egghead will survive. What's important is that in this case the company realized that it needed to do business in a radically different way and took action.

2. Can any of your existing product line be sold online, and will this provide additional benefits for your customers?

Over the years I have tried to determine what distinguishes outstanding, highly profitable companies from their competition. One answer is that highly successful companies are "fanatic" when it comes to focusing on how they can best serve their customers. This devotion includes ensuring that top officials meet with customers, to carefully listen to customers for insights into their needs.

This same customer-oriented focus is critical to building a successful online sales effort. I find that outstanding companies don't say, "Great! Now we can sell online." Instead, the instant reaction of their executives is "Great, this will help us serve our customers better because . . ." and "We need to be certain that we include this feature because our customers have been telling us that . . ." Here are a few examples:

Initially, Sears Roebuck had an information-only Web site. In its first foray into online sales, the company experimented at two levels:

First, it introduced 3,500 of its Craftsman tools available for purchase online, testing the extent to which this homegrown brand, as opposed to the other name brands, drives sales. I suspect this choice reflects Sears's concern that at some point more and more leading brands will sell direct to consumers. Sears started the valuable process of determining how it can use its own brands to serve customer needs.

Second, because it is a focused site, Sears has surrounded the product with appropriate information, including detailed instructions on home improvement projects ranging from soldering a copper water pipe to building a trophy case. This additional information represents real added value: Buyers may want assistance with specific products, and Craftsman owners will be able to access this information evenings and weekends when they decide to start a project.

For Ticketmaster, the Web has improved the manner in which tickets can now be sold. One of the most vexing issues for Ticketmaster customers is their difficulty in assessing the likely view from seats purchased over the telephone. For most venues, Ticketmaster customers on the Web can now see a seating chart that indicates what the view of a specific stage is from the seats they are thinking of buying, offering a greatly improved service to customers.

JEM Computers specializes in selling excess manufacturers' inventory and refurbished PCs. After selling by catalog, JEM turned to the Web when the company realized that through online sales it could improve its offerings: Since the company's specialty is overstocked and refurbished PCs, the Web's timeliness actually meant an enhanced product and higher value for customers. Instead of finding out about a refurbished PC in three months when a new catalog came out, customers could buy the most up-to-date machine possible, simply because the lead time for notification of availability was shortened.

3. Does the Internet create an opportunity to enhance your existing product and build additional revenue streams?

The communications capabilities provided by the Internet may create an opportunity for you to continue to offer the same core product but in an enhanced way. These enhancements might involve *lower pricing* as a result of automation, *faster turnaround* on service applications, *better customer service,* or *more useful products.*

One striking example of enhancing an existing product involves Simon & Schuster, which prints one in every five college textbooks. The company is using the Web to enhance the value of its product. Through a Web-based service called NewsLink, the company updates 50,000 to 100,000 pages on a daily basis, thereby providing students with information that is current. The company includes a unique identification number on each book, and students can access a news feed throughout the term. If the book is subsequently sold to someone else, the buyer can reinstate the news feed for $10 to $15 per term. This clearly represents a win-win melding of traditional print textbooks with the evolving capabilities of the Web. Students receive more relevant learning materials. At the same time, this creates a new revenue stream for Simon & Schuster when students purchase used books.

4. Does the Internet create an opportunity for you to develop entirely new products or services based on your core competency?

While digital cameras and the resulting pictures have not yet replaced snapshots in the family photo album, Kodak foresees a day when photos online will be faster, cheaper, and better. Kodak bought a controlling interest in PictureVision, a pioneering company that has streamlined the process of getting photos online by providing photo finishers with software that lets them upload customers' images to a Web-based archive accessible with a personal code.

The greatest advantage offered by Web-based photos is access. Now, instead of storing photos in albums or old shoeboxes, families can upload the pictures (or for people working with traditional film and cameras, have the photo finisher upload them), and extended family, including grandparents in Florida, can view them as easily as family members can.

Even accounting firms are finding ways to develop new offerings via the Net. PricewaterhouseCoopers is selling its Web visitors information it has never made publicly available before; the annual $540 it collects from each subscriber to the Tax News Network is the beginning of a new revenue source.

Meanwhile, Ernst & Young, a competitor, has created a different kind of new service called Ernie. As one of its offerings, this new service allows subscribers, who pay a flat fee, to submit a specific number of questions online. The questions are routed to a knowledgeable member of the worldwide firm, and the subscriber receives an answer from an expert at a fraction of the usual cost.

For radiologists who spend much of their time at workstations reading the output of digital imaging technologies, what could make more sense than a medical journal for radiologists that is published totally online? The Radiological Society of North America is pushing the envelope of academic publishing with a peer-reviewed publication that exists only on the Web: RSNA Electronic Journal. The journal offers turnaround time of as little as a month, compared with the six to nine months it takes a traditional journal to move a paper from submission to publication. The journal can be distributed faster and at lower cost, and it's a better product because it's more timely. Like encyclopedias, scholarly online journals—particularly those in science—are destined to migrate here. (Faster, easier, better, cheaper!)

Finally, it's always important to remember that if there is the potential for an enhanced product, if you don't act on it, someone else will—even if that someone else lacks your roots in the business.

The examples presented above vividly demonstrate the double-edged sword of the Internet. On one side, it presents tremendous challenges for businesses and may threaten their very survival. On the other side, it creates opportunity. Using the Internet, companies can take the skills and resources they have and build new profitable businesses.

FINDING NICHE MARKETS
FOR SPECIALTY PRODUCTS

If you've always had a great product but have had a difficult time finding the right audience, the Web may be perfect for expanding your business. Specialty products have a natural place on the Web. The owner of an Italian food specialty store in New York City found that while New Yorkers can buy salami anywhere and do, people around the country were delighted to know where to buy salami; his product is now shipped to many new customers nationwide.

ONLINE PRICING STRATEGIES

Success on the Web also dictates certain important principles regarding pricing strategy. These are discussed in depth in Strategy 7.

THE FUTURE OF THE PRODUCT

Schoolchildren are often taught "the best defense is a good offense," so being proactive is a vital competitive strategy. At the same time, I opened this chapter by noting that the most successful companies are obsessed with listening to their customers. As will be discussed in Strategy 7, many companies have learned a great deal about what their customers need and about future demand by carefully watching how their Web sites are accessed. This works even when the site is accessed anonymously: Companies can learn a great deal by asking themselves why certain areas of their sites are or are not receiving frequent visits.

THE BATTLE PLAN

Quality is always a critical component of success. The Internet does not change this fundamental rule. Nonetheless, companies can turn to the Web to expand their product reach, enhance the product itself, and improve customer service. If your compass always points to improving the customer experience, then sorting out how to use the new capabilities of the Web to enhance your product or service becomes straightforward and may even start to seem easy.

A critical question to ask is "Can new Web-based technologies enhance my product or service, and how will they improve how I am serving my customer?" The cost and effort involved in any Web initiative you're considering should always be weighed against specific, anticipated benefits.

In analyzing a potential initiative, it is also important to keep in mind that the competitive environment is not static: You need to consider what will happen if you don't take a certain action and your competitors do. In a hypercompetitive world, you can't assume ongoing customer loyalty, you need to keep moving forward to earn it.

Get Personal

I t is one of the supreme ironies of the Internet that the computer, so long derided as impersonal, is now being used to create highly personal experiences for Web site visitors. Because a computer can sift through vast amounts of existing information according to preprogrammed rules, computers can now take company information (or special interest information) and combine it with information supplied by prospective customers and digest it in a way that is meaningful to each individual.

This chapter introduces the best ways businesses are finding to personalize selling and customize products in order to build business. Although many of these methods are still in their infancy, a great deal can be gleaned by learning what leading-edge companies are doing currently and what technology is likely to make possible in the future.

To attract new customers, companies can now establish Web sites that provide highly individualized recommendations based on information provided by the customer. In this way, it's possible to tell a prospective customer exactly which product, among a plethora of possibilities, is just right for that individual or that business. The strategic idea behind these efforts is clear: By empowering the customer with information about how a specific offering meets his or her needs, the company positions itself as a knowledgeable place to buy and also demonstrates how its products are just right for that individual customer.

This information, properly used, gives an existing supplier or retailer a clear leg up on the competition: The company is able to use its information to establish itself as the supplier of choice and to suggest services to existing customers before they request them. For example, companies such as Bay Networks, a West Coast–based technology firm, have started to maintain what are effectively private Web sites for individual clients so that they can both provide extraordinary service and recommend new products that will be of value to these customers.

In the past, this type of personalized communication was virtually impossible: Retailers and suppliers lacked the detailed information necessary to provide these recommendations and the ability to cost-effectively communicate it to individual customers.

However, today we have moved into an era where technology allows firms to cater to the individual needs of customers in a way that has not been possible before. The value of these initiatives, both in tightening the bonds with existing customers and in attracting new customers, is high.

Another business tool made possible by the Net is something that previously was available only to the wealthy: product customization, done quickly and inexpensively.

We can start by considering something simple such as the customization of greeting cards. While the thought of being able to create a uniquely personal online greeting card—complete with a personal photograph—sounds nice, things get a lot more interesting when you think of the possibility of being able to "build" your own CD online, putting on it only songs that are your personal favorites. (Today that CD will be mailed to you; tomorrow you'll download it from the Web right after you make it.)

The textbook publishing business has found that customization via the Internet is changing the way it does business: McGraw

Hill's Primis Publishing has established a Web site where college professors can "build" their own textbooks. By going online, an instructor can select from 12,000 documents ranging from history-making legal cases to articles about chemistry to put together a book that is tailor-made for his course. Once the chapters are chosen, the textbook can be printed and ready for distribution in about six weeks. Costs run from about $15 to $100, compared with about $50 for a standard textbook. Based on this information, it comes as no surprise that the growth of custom publishing of college texts is outpacing traditional textbook publishing by 10 to 15 percent (15 to 20 percent versus 5 percent).

In an era where a single click can take customers away, leading-edge companies are finding that ongoing, personal relationships can be key to winning HyperWars. Personal selling, "marketing intimacy," if you will, deepens the commercial relationship, adding tremendous value for the customer ("This saves me so much time—and it's just right for me!") and making it painful and costly for the customer to leave ("Why should I buy from anyone else? Company X knows exactly what I like and need"). Therefore, the more your company can "get personal," the more likely the potential for long-term retention of customers.

To help your company benefit from the potential of personalized selling, this chapter will demonstrate how companies are making use of several levels of "personal" recommendations. We'll also highlight some developments in customization and personalized selling via e-mail that are in their infancy but that will soon be the norm.

In adopting any strategy based on personalization, privacy is a central issue. The ability to make the customer feel comfortable enough to give you the information you need to create a tailored product or solution is among the primary challenges companies must overcome. I'll share several different ways to approach this issue.

THE UNIQUE POWER OF THE WEB
TO GET PERSONAL

The Web's ability to help companies establish marketing intimacy is possible because a Web site combines five elements:

1. An opportunity to present information in an interactive format, permitting customers to express their preferences to the owner of the Web site
2. Virtually costless online communications between the potential buyer and seller
3. A visual component that lets potential buyers see prospective purchases and how they would appear if personalized
4. The ability to store tremendous amounts of personal information about their customers
5. An unprecedented ability to create systems that configure products so that costly errors are eliminated, thus reducing the expense of creating custom products

Presently, one-to-one selling can be viewed in two general categories, recommending and customization.

THE RECOMMENDING FUNCTION

Brick-and-mortar companies have long known that for the most part, either a motivated buyer has done a great deal of research on a particular product, or a product has been enthusiastically recommended by a friend or business associate. This knowledge leads to two keys to establishing a Web site that motivates buying:

1. Ample information (so that the buyer needn't do any more research—and, in the process, possibly depart your Web site for another)
2. A positive recommendation

The Value of Recommendation Systems to Consumers

While in the past, the most effective recommendations were generally person to person, there are several reasons why these technology-based recommendations are valued now. One has to do with the dizzying array of choices that now faces consumers. A cosmetics firm may offer hundreds of shades of lipstick or eye shadow; financial services companies have become true supermarkets filled with choices; well-known cold remedies now have multiple versions that end with words such as "plus" and "sinus" and "cough."

A generation ago, the dilemma posed by the array of choices was generally solved through one-on-one interaction (selling, if you will). Someone with a cold would have stopped by his or her local pharmacy and chatted with the pharmacist about which of the three or four cold remedies carried by the pharmacy would be best.

Today, the pharmacist is hard to find (and is often employed by a separate entity within a chain drugstore), and the number of choices of what to take for the common cold is mind-boggling. Do you want to take the medicine in the day or night? Do you need an expectorant? A cough suppressant? Decongestant? Antihistamine? Something for fever and chills? Normal strength or extra? Customers could get a headache simply trying to decide! Today cold or allergy sufferers need only click to the Warner-Lambert "allergy-cold" site, where they are asked to click off their symptoms (itchy watery eyes, runny nose, cough, etc.). The site then serves as a "friendly pharmacist" and recommends the product that will be right for them. While consumers recognize there is a given bias to any company-sponsored Web site, they still welcome the guidance because they trust the branded product provider and they are overwhelmed by choices.

Depending on your product and customer, several types of recommending functions may work best. But first, let's take a moment to consider how to judge effectiveness.

Recommendations Based on General Information
Provided by a Customer

"I've just met you,
but I have a hunch I know what you'd like."

As a result of my work, I have analyzed a large number of database marketing initiatives, and one online company is implementing, without question, the best strategy for using customer data to increase sales that I have ever seen.

From the outset, Amazon.com has been a leader in understanding how to use the Web and associated technologies to build its business. When you choose a book, the Amazon site will automatically indicate that "customers who bought this book also bought . . ." and list the related titles based on the purchase patterns of other customers.

This form of recommending, combined with a customer database, is flat-out brilliant. First, since Amazon is drawing on the experience of its total customer base, this effort raises absolutely no privacy concerns. Second, the most effective recommendations are based on how people actually behaved (i.e., how they actually spent their money), as opposed to what one might think they should be interested in. The method is more knowledgeable than even the most diligent clerk jumping in and saying, "Well, if you like that, then you're going to love this!"

In addition, the nature of Amazon's products lends itself to encouraging a customer to buy more at the moment of checkout. If a prospective customer is buying a nonfiction book, he or she is likely to welcome additional suggestions for researching this topic. This capability serves as a database-powered means of intelligently cross-selling products.

I've often wondered how much this use of the customer database has raised Amazon's sales. However, my educated guess is that it is quite a significant amount.

I would suggest that any business looking to effectively compete in today's hard-fought environment take a careful look at what Amazon is doing to derive value from its database. It may not directly apply to your company and product, but there may be a creative new way to examine your own possibilities by considering whether there is an equivalent way you could use your sales data to effectively cross-sell to consumers.

The Friendly Clerk Asks Some Questions (Educated Guesses Based on Personal, Anonymous Feedback)

> *"Let me tell you about this product
> and recommend what seems right for you."*

Thanks to the interactive aspect of the Web, many technological rec-ommendations are firmly based on consumer feedback. For exam-ple, at Clinique, the Web site employs a method that is very similar to a system employed at Clinique's retail counters. A Web site (or in-store) visitor provides information on hair, eye, and skin color as well as information concerning breakouts, response to the sun, and signs of aging. Based on this information, specific products are recommended. (At some point soon, this system will seem very prim-itive, because before long, the technology will be in place for cus-tomers to scan in photographs of themselves so that recommendations can be even more personal by allowing prospects to "see" how they would look using different products.)

In the travel industry, Club Med is a business that realizes that attracting a customer's interest is one thing; securing the sale is another. Although many people are positively inclined to take a Club Med vacation, they aren't certain where they "belong." A friend went to a Club Med in Mexico; a neighbor swears by the Florida Club Med. How to know where to go? Stepping in to serve as the friendly travel agent is the Club Med Web site. In an area called "My Ideal Village," visitors are asked to specify their needs, including the type of club (clubs with family facilities, clubs for adults over eighteen, etc.), the type of activities they are looking for, and the price range they want. Based on this information, the site does an instant search and comes back recommending certain Club Med sites. The site then provides photos and extensive details about the recommended clubs. Customers can then request reservations or rates through online e-mail at the site. In addition, Club Med provides an 800 number so that the customer can have the reassurance of booking through a real person.

Today 40 percent of all adults buy at least one toy during the year, and many of these toys are purchased as gifts by adults who don't have children of their own. eToys, an online toy store, offers personalized services that create a shopping experience that is superior to those offered by traditional businesses. Instead of leaving clueless customers fighting their way through brick-and-mortar aisles with real, rambunctious kids, eToys offers the home-based customer a choice from a sampling of toy suggestions based on age and gender. The eToys "recommending" system is like a friendly clerk walking you through the store: "I know this age, and I've seen what they whine for. Try it, you can't go wrong, and I'll wrap and send it for you." What's more, the customer can choose recommendations that are developmentally appropriate and can be geared toward being educational; as a result, the buyer actually feels virtuous. Who can resist?

Customers can select ways to broaden and then narrow their search. They can browse the virtual aisles by specifying their interests: dolls, puzzles, or award-winning toys, for example. Customers of eToys can also enroll in its reminder service, which extends the convenience of the store by ensuring that customers don't miss gift-giving dates. The service will send the customer an e-mail when an important gift-giving date nears.

To the public, there's little that's more confusing than figuring out what to do with one's money, so, of course, information and recommendations are vital in this area. Financial service companies are having a heyday with the possibilities offered to them via the Web.

At John Hancock's MarketPlace, visitors can learn about various financial services, get general financial advice based on some anonymous profile information ("Are you recently married? Are you a recent parent? Are you looking to save for retirement?), learn about various financial products, or actually buy direct from the life insurance company. At Met Life, a similar setup exists. If a visitor registers with the "Metgician" (Snoopy dressed as a magician) and answers eight relatively general questions (age, gender, marital status, professional life, state, vehicles owned, etc.), the "Personal Met" will make some recommendations on topics ranging from retirement planning and homeowners insurance to home improvement costs and child care. Visitors can use the system without identifying themselves, but Met Life invites them to provide personal information by allowing registered visitors to store information about themselves and their needs so that they can return to the site for additional information without having to start over.

Although Met Life provides an incentive for visitors to register, all of the above examples share an important characteristic: They provide valuable recommendations to consumers without asking the name of the prospective customer.

If the visitor is willing to give a little more information, then the recommendation can be customized further and will be that much better.

Some Criteria for Effectiveness

Personalization, like any business initiative, should be used judiciously. There are real costs that accompany developing personalized applications: They range from the hard cost of development dollars to the cost of disappointing customers or prospects who expected a better experience. It's important to assess whether a personalization effort will be effective and contribute meaningfully to new sales or customer retention before investing the time and money in bringing it to market.

The criteria for establishing an effective recommendation system are different for every product. The system chosen is dependent upon a balance of the following: (1) the nature of the product, (2) the amount of information needed from the potential buyer to make an effective recommendation, and (3) the likely willingness of the prospective buyer to share this needed information with you.

As you go through this process, be aware that we have entered an age in which consumers *want* to educate themselves about everything from which refrigerator to buy to how to get the best medical treatment. They no longer trust intermediaries—advice from dealers, brokers, agents, and even health care providers is being checked and rechecked by today's consumer. A well-designed Web site helps meet this need, and the information should be as rich and informative as if the consumer were meeting with one of your top salespeople.

Wait a minute? Am I saying that consumers will trust information they receive from companies on the Web as much as (or more than) they trust sales representatives? All evidence to date suggests the answer is yes. People believe that no trusted brand will risk its good name by posting misleading information on the Web.

As you go about creating an online recommendation system, your overriding goal should be to make the online shopping experience better than what a customer might encounter in the physical world. Therefore, you need to ask yourself: "What can my company accomplish using this medium that can't be accomplished in the physical world?"

The Central Issue of Privacy

When I developed prototypes of my newsletter, *Bruce Judson's Grow Your Profits*, I asked friends and family members to take a look at it. The immediate response from my mother-in-law (one of a growing number of savvy senior Internet users) was "How do I know you won't sell information about me?" This vividly demonstrated to me that privacy issues have now replaced credit card security issues as the number one impediment to building an online business.

As discussed earlier, the primary feature that distinguished the different types of recommending activities discussed above is how much information the consumer will share with the business running the Web site. A central issue for businesses is how to make prospects comfortable with sharing enough information about themselves that the company can provide valuable recommendations.

Here are a few ideas on this topic:

Consider gathering data anonymously. Recommendation systems that provide advice to consumers anonymously are likely to be far more popular and, therefore, more effective. In these cases, you are able to provide the prospective buyer with value without needing to jump the hurdle of "you can trust me with this information."

If you'd prefer to record the identities of site visitors, you are one step ahead if you are a recognized brand. Consumers are already predisposed to believe in your company name. In this case, address the privacy issue head-on with a statement that precedes your online registration form: "The information gathered here is to help us better serve you; we will not share or sell this information to anyone."

If you are not a well-known brand, a strong privacy statement is recommended, but you should likely take other measures to more fully establish your legitimacy. There are now a number of entities, such as the Better Business Bureau's BBBOnline, that have developed certification programs. These programs warrant to site visitors that you are a legitimate business and agree to follow certain ethical business practices. Members of BBBOnline have the right to prominently display a logo on their home page. I strongly advise businesses to look into this program.

Finally, if you plan to share customer data with others, then you need to be explicit about it and receive the customer's permission. The worst thing you can do is fail to inform the customer that some piece of information you determine about him or her may be used in some way the customer does not expect.

CUSTOMIZATION

Product customization is possible because of a convergence of two elements: the Web as a one-to-one communications technology and manufacturing processes that allow for the development and delivery of custom products.

Designing your own custom desktop computer at Web sites such as Dell, Gateway, and Apple has become "commonplace," but what about designing your own swimsuit or, perhaps, golf clubs? These customized services provide several valuable benefits for companies: (1) They help to drive new business, and (2) customers are willing to pay a premium price for an individually designed product, which means these products and services sometimes have higher profit margins than standard offerings.

Calla Bay's Web site lets women design and buy their own swimsuits online. Shoppers can mix and match fabrics, colors, and styles to create their own suits. As the suit is created, the visitor can view the suit on a 3-D model and then various fabric possibilities can be considered. Once the visitor has selected the perfect combination, the suit is made to order and shipped to her home in about two weeks.

Chip Shot Golf has watched company sales more than double, with custom-club ordering making up 80 percent of the sales. Visitors to the site can select clubs based on choice of grips, shafts, and club heads, and tailored to an individual's height, hand size, and playing style.

Standard & Poor's, a company that has primarily served Wall Street and the financial community, is using the Internet to expand its business into consumer services. Like other financial service Web sites, S&P offers general financial planning advice based on answers to an online questionnaire. However, the company has created a unique way to expand its "recommending" into customization. Customers who pay a $9.95 per month subscription fee are entitled to personalized financial planning recommendations, written by registered investment advisors. This represents an entirely new source of revenues for S&P.

For every nutrition-conscious customer who is tired of unbottling and taking ten or twelve vitamins for health, Acumin Corp. of Malvern, Pennsylvania, has established a Web site that permits visitors to create personally tailored vitamins. After the visitor answers a questionnaire regarding nutrition, exercise, lifestyle, and medical history, a nutritional supplement is recommended and can be custom manufactured for purchase by each customer.

The ability to interact with customers and say, "I can create the product that is just right for you," is one of the most powerful features of the Web. Expansion of this Internet capability will be the inevitable result of the combination of (1) manufacturers creating products that permit increasing customization, (2) further advances in computing power at ever decreasing costs, and (3) new software that is continuously enhancing the options available to businesses to personalize.

As we go to press, the Gap includes a "get dressed" interactive feature that lets visitors put clothes on a generic man or woman, and cosmetics sites show how certain basic "faces" will look with certain colors. However, a CD-ROM from *Cosmopolitan* already exists that allows users to scan their own pictures into the computer and then give themselves makeovers. From there, it's a small jump to imagine online sites that allow you to scan your own picture into a particular format, and then see how you look in certain clothes, special makeup, or with a new hairstyle.

What Forward-Thinking Companies Are Doing

Smart companies realize that the more involved the relationship becomes with the customer, the better their rate of retention. They are implementing this strategy in a wide variety of creative ways, involving both personalization and customization, and will certainly enhance their applications significantly over time:

1. Companies are creating extraordinary convenience for customers and establishing systems to speed the fulfillment of custom orders.

Ford's intranet permits custom ordering and delivery on every car or truck. Dealer and customer can sit at a PC, specify the exact configuration of look and accessories the customer wants, view it on the monitor, and get a confirmed delivery date. Currently this better, more personal service is available only through dealers but one day will likely be available in our own homes. What's more, where in 1996 it took more than fifty days to get the Mustang of your choice delivered from the plant to the dealer, today you'll get that Mustang in fifteen days—Ford's goal is to manufacture the majority of its vehicles on a demand basis by the end of 1999, with delivery in less than two weeks after the order. This would also save billions of dollars in current inventory and fixed costs. (Read more about savings on inventory in Strategy 3.)

2. E-mail communications and reminder systems are going to be used increasingly.

The ultimate goal of every business is to have a customer who *wants* to hear from the business about new products that he or she might want to buy. E-mail, for the first time, provides this type of powerful tool, since it's an almost costless communications vehicle. At American Greetings, members can store important dates of friends and relatives and receive e-mail reminders. The consumer receives a valuable service, and American Greetings looks to sell a card at that moment by providing extraordinary convenience.

3. In business-to-business selling, smart companies are also linking tightly to their customers.

Ross Controls, a maker of pneumatic valves—devices that force air through machine tools—owes its continued success to mass customization. Customers become part of the design process rather than passive recipients of a standard product by using state-of-the-art computer-based engineering technology to create customized pneumatic system solutions within days. While customers currently must meet with a company representative to work on the design of what they need, there is little doubt that down the line this could all be accomplished via the Web.

A few leading-edge technology companies are offering custom password-protected Web sites for their corporate accounts and high-volume small business accounts. These sites are typically designed to simplify the buying process, and they offer (1) customer online malls offering products preselected by the company to be bought by employees at volume discount prices, (2) electronic mail links to account managers responsible for serving that customer, (3) the ability to track the status of orders, and (4) dramatic increases in the speed of order fulfillment through the elimination of paper forms and a reduction in errors.

I anticipate that winning companies will increasingly offer this type of specialized service. It is a clear way of providing customers with additional value and of tying the customer more closely to your company. Most companies will also see an increase in revenues from these accounts.

"But What If They Write???!!!"

When companies first began establishing Web sites, most feared they would receive an overwhelming amount of e-mail that they wouldn't be able to handle. In many cases, two unproductive strategies emerged. Some companies designed Web sites that did not permit the visitor to contact them, making the site little more than a brochure; other companies permitted e-mail messages and then often didn't respond to them. In either case, this did not leave potential customers with any feeling that their visit "mattered" to the company: the exact opposite of the tight relationship a company wants to build between itself and its customers in a HyperWars environment.

What was called for, of course, was a new category of employee: "e-mail respondents"; yet no one wanted to add staff at an additional expense to a project that was in its infancy. Enter "response software."

Several intelligent software systems, such as Brightware and ResponseNow, are now available to handle and route questions, so there is no excuse for underserving your customers with the first and most basic customer service—answering their questions.

ResponseNow has created a product designed to facilitate response to and management of customer e-mail quickly, personally, and efficiently. It is a virtual call center that enables an unprecedented level of interaction between a company and its Internet customers, automatically routing inbound e-mail to the correct representatives. The software manages, tracks, analyzes, and archives the entire life cycle of all customer e-mail correspondence.

Brightware has created a system that mimics a talented salesperson: It answers customer questions, sends information, makes product recommendations, and refers potential buyers to sales representatives who can close a deal. Brightware works by relying on knowledge-based technology to look for preselected keywords before sending the user a personalized response. Its software can automate more than 85 percent of the inquiries businesses receive over the Web, and the system is accurate 98 percent of the time.

Nike uses Millennium's Echomail to manage the 2,000 messages it receives each day. The software can "read" messages, ranging from "Send me a catalog" to questions involving where to look for particularly narrow shoes.

Companies who are successfully managing their e-mail via electronic means are also taking the opportunity to benefit from the knowledge they glean. At Aptex, if someone sends an e-mail requesting a brochure, Aptex software can read the e-mail, understand the request, send the document, and then update that person's profile based on the product request. Aptex provides Charles Schwab and Suretrade.com the solution to the problem of receiving about 5,000 customer e-mails a day. (It used to take Suretrade seven days to respond to its e-mail; now it takes only fifteen minutes.) Since most of the queries revolve around similar issues ("How do I open a brokerage account?" "How do I fund my account?"), it's relatively simple to come up with an automated response. Aptex's technology can read the e-mail, and in the case of Suretrade, automatically respond.

The online Direct StockMarket runs a small online market for new high-tech companies that want to sell shares directly to individual investors. The ErgoTech's E-mailroom software routes inquiries directly to the staff person responsible for that stock.

Companies are finally transforming their Web sites from marketing brochures to vehicles that turn visits into sales leads. Those who develop a Web presence and fail to respond to electronic inquiries in a timely manner run the risk of losing existing and new customers. Industry research also shows that only 30 percent of Fortune 500 companies respond to questions directed to these companies through their Web sites, which means a large number of customers aren't getting the personal attention winning companies need to provide.

A Proactive Combatant in HyperWar

An excellent example of how personalization or customization can save your business is demonstrated by a printing company that could have gone under but fought to stay on top:

Since Ford Motor Company established its intranet, the company has had the opportunity to reach out to businesses across the nation, seeking the best services and the best prices; no longer was it easier to do business with local suppliers.

One of the businesses affected was a former Ford subsidiary, Geometrics Results, Inc. (now owned by MSX International) that had formerly supplied Ford's nearby Dearborn, Michigan–based product-development facility with everything from printing and document management to building prototypes. MSX's print shop depended on Ford for 90 percent of its business—it had Ford letterhead on its shelves and had memorized deadline schedules for each department. The staff knew Ford's needs inside and out.

Realizing that Ford might find other suppliers over the Web, MSX took action. The company obtained permission from Ford to install a software program that reflected MSX's intimate knowledge of the way each person at Ford bought a print order. Now, using customized software created by MSX, a Ford employee can go through a series of prompts and attach text and art files to be sent seamlessly to MSX for printing. Each e-mail is carefully monitored by MSX to be certain that there are no problems, and all is done electronically, including proofs. The method created is reportedly more efficient than the one used previously.

MSX's smart response to the breakdown of geographic barriers and its embrace of customization has let it retain Ford's business, and it has opened up new markets as well, as MSX begins to create these specialized systems for other companies.

The Battle Plan

As you consider the possibilities offered by personalization and customization, I suggest that these initiatives be weighed against these criteria:

1. Will the initiative enhance relationships with my customers by adding convenience or a better ability to meet their needs?
2. Will the initiative result in potentially high cost-savings for my organization (as in the case of an intelligent product configurator that eliminates costly order errors)?

3. Will the initiative serve to grow my business, either because of its impact on my existing customers or because it will enable my business to cost-effectively serve new markets?

With these questions firmly in mind, you can evaluate how you would like to proceed. Do the costs involved meet at least one of the criteria listed above to make the payoff worthwhile? After resources have been allocated, and you move from planning to development, new questions about features and capabilities will inevitably arise, so keep these criteria as a reference throughout the development process as well. They will again provide a guide to what features are worth the cost involved.

Create the "Total Solution" for Your Customer

Y ou don't need to shop anywhere else" is the message from any company trying to provide the "total solution" for customers. This is particularly important in an era of heightened competition, price wars, complex products that must work together, and click-of-the-mouse window shopping. By creating tight bonds with customers, you may be able to prevent a good number of them from going to your competitors. Consider this case of a company that found a way to offer the total solution via the Web:

One of the most unlikely companies for doing business on the Web must surely be an enterprise selling custom blinds and shades. In the brick-and-mortar world, a customer would visit the retail store, select the style of window covering desired, and then set up an appointment for a store representative to come to take window measurements. A few weeks later, custom blinds (or shades) would be delivered and installed by the store.

Yet No Brainer Blinds and Shades, based in Houston, Texas, has found that by providing the "total solution" for customers, it has been able to make quite a success of selling via the Web. For this entrepreneurial business, the "total solution" involved four components: low prices, fast delivery, high-quality information and design ideas, and a wide selection of products.

Low overhead for selling from a Web site helped No Brainer accomplish its first goal—low prices. Then, because owner Jay Steinfeld and his wife ran a brick-and-mortar custom blind business, they were able to talk to their contacts about the next vital ingredient for strong sales: fast delivery. Vendors promised them the same one-day delivery policy the owners had with their brick-and-mortar business.

The next challenge was customer information. Because they were aiming at the do-it-yourself market, Steinfeld felt that getting adequate information about how to measure properly and how to hang blinds would overcome the one remaining obstacle to blinds via the Internet. They worked hard to make the site visually interesting, user-friendly, anticipatory, and so information-intensive that even a first-time blind buyer could do it. The site features quick-time movies demonstrating how to hang blinds as well as additional decorating ideas. It's also done with good humor—audio testimonials are offered by Mae West and Abraham Lincoln as well as real customers.

Steinfeld cites two specific reasons for creating an information-rich Web site: First, it establishes his credibility with potential customers—visitors to the site can see his design expertise. Second, Steinfeld believes that if you give something free to people (in this case, valuable information), and they know that no strings are attached, then they are more likely to buy from you.

As the No Brainer site began to get customer feedback, Steinfeld learned that the "total solution" needed to include a wide variety of colors. Although the majority of a customer's purchase might be white or off-white shades, he notes: "If I didn't have the one pink shade for the formal bathroom, I was likely to lose the sale." Customers clearly wanted one entity that would handle all their needs. In sum, No Brainer proves that all sorts of products can be moved over the Internet—by providing the "total solution."

Because online customers can "search the globe" for the product they want at the price they prefer, it is vital that businesses with an Internet presence determine a way to differentiate themselves from the competition. Many companies, both those that sell to individuals and those that sell to businesses, will find that part of this distinction will involve offering the maximum in customer convenience—providing a "total solution."

In the physical world, this is generally thought of as "one-stop shopping." That same notion holds true for the Web, but its implementation now often extends even further. Ultimately, companies will cease to be "product" companies or "service" companies but will instead become "customer solution providers"—entities that mix products and services in the pursuit of offering solutions. Just as No Brainer Blinds and Shades went far beyond being an "order taker" for window treatments, other companies—including large corporate entities—are finding that on the Web their products need to be cocooned in information and services that provide customer value.

At times you may find that it's impossible to "be all things to all people," but as you'll see, here, too, the Web offers a special opportunity through partnering. Many companies are realizing that it's hard enough to keep up with their core business, let alone add something new. Yet through strategic alliances, companies are finding ways to link up with complementary service providers to provide a seamless one-stop-shopping experience for their customers.

This chapter will illustrate that hypercompetitive businesses can partially fend off competition by being "so good" that their customers never feel they need to look elsewhere. I'll start with some of the simpler ways to provide customer satisfaction and move forward to more sophisticated implementation of the idea.

PROVIDING THE "TOTAL SOLUTION"

Companies that dedicate themselves to providing the "total solution" for customers generally create new opportunities to counter competitive threats and find that they win market share. These companies generally have three attributes:

1. They define their business broadly.

In a sense, they are quick to recognize that a particular product may, in the near future, have limited appeal, but that other related services will be needed. Often, these services will draw on the same core capabilities as the first product and will be a natural extension of the brand for customers.

2. They are not afraid to partner.

The notion of working closely with other companies is one that appears frequently in this chapter. Rarely can one company do everything well, but one company can successfully manage a process that involves multiple entities well. This approach is the exact opposite of a "not invented here" mentality.

3. They have a strong commitment to customer service.

This is crucial, and it must be more than a "lip service" commitment. Companies that convince customers to follow them into a brave new world must be absolutely committed to making this new world work for customers. Customer loyalty lives or dies on how the customer experiences the product because of the service offered.

Companies that can most completely and easily (i.e., most conveniently) meet the needs of potential customers will attract new customers, win customer loyalty, and thereby increase their market share.

MAKING IT EASY FOR CUSTOMERS
TO DO BUSINESS WITH YOU

For businesses, one of the problems inherent in the Web is the ease with which consumers and businesses can move from one company to another in search of lower prices.

Sophisticated companies have recognized that the "total solution" for customers may often involve creating new, enhanced services. These added services deepen the customer-company relationship, enabling the vendor to offer more value to the customer because of the intimate knowledge it has acquired of the customer's needs. And as a result, the customer feels an increased loyalty.

These dynamics are no different in the business-to-business arena. The longer the relationship, the better the "seller" (or "solution provider") knows the needs of the customer; and the better the service, the less likely the customer is to shift its business elsewhere.

It seems that almost every day we read or hear that we live in an increasingly mobile society. For any company, it's always more difficult to attract new customers than it is to maintain current ones. The Internet offers a unique opportunity for a brick-and-mortar business to remain available to customers, wherever they might be.

One example of this is Royal Bank of Canada, which faced a potential of lost business because of Canadian "snowbirds" — Canadian natives who prefer to winter in the United States. When Royal Bank had the opportunity to acquire the Atlanta-based Internet banking arm of Security First Network Bank, the first bank to exclusively sell its services over the Net, it took it. Company officials saw it as a way to continue to provide services to their customers, and it was a perfect way to keep an American camel from getting its nose under the Canadian tent. By acquiring an Internet bank, Royal Bank can interact with the "snowbirds" wherever they go, and there is no need to acquire a more costly brick-and-mortar U.S. base.

In addition to Amazon's ability to "personalize" the shopping experience (see Strategy 5), the virtual bookseller offers the ultimate in customer convenience: Addresses and credit card numbers of regular shoppers are kept on file, meaning that book-buyer regulars need do nothing but click on their purchases before signing off line. Amazon's account maintenance system provides its customers with secure, password-protected access to everything about their account at anytime. They can view their entire purchase history, track orders, customize virtually everything about the system to their own tastes, and reach the bookseller by e-mail at almost every turn. Orders are confirmed by e-mail, and mistakes are corrected completely at Amazon's expense via a pre-addressed padded shipping bag mailed along with the correct book.

Sporting goods retailer Recreational Equipment Inc. (REI) has found that round-the-clock access is well worth the effort. REI generates 35 percent of its online orders from 10 P.M. to 7 A.M., when no REI stores are open and no mail-order operator is available.

Or what about a simple issue such as language? If customers can read in their native language about a product or a service, they are more likely to have the confidence to buy. As a result, more and more companies are adding multiple languages to their Web sites in order to appeal internationally. AMP's online catalog of electronics products supports eight languages; a visitor simply selects the language of choice when he or she enters the site, and the rest of the transaction is conducted in the preferred language. While translation adds to the cost of doing business, more and more machine translation systems are being developed that are less expensive than human translators.

PARTNERING TO PROVIDE
ONE-STOP SHOPPING

The appeal of one-stop shopping for consumers as well as in business-to-business transactions is universal. It's convenient for the customer and it lets the merchant maximize a sale. The more companies can do to prevent customers from having to go elsewhere, the more loyalty they will build.

Any business can take two important steps toward fulfilling a customer's one-stop-shopping needs in the online arena:

1. Decide on the core product or service that you are offering for sale.

2. Make a list of all the products and services a customer must possess in order to take advantage of this core service. By extending your product and/or service line, you offer greater satisfaction to the customer and potentially increase profits through add-on products (through traditional cross-sales or via a sales fee from an alliance partner).

To provide a "total solution," your goal should be to provide the customer with every item on that list. However, "provide" does not necessarily mean "sell."

One of the advantages of the Internet is that it is relatively easy to create alliances with other businesses in such a way that both your customer (by being able to accomplish one-stop shopping at your site) and you (by being able to satisfy the customer) benefit. The key is to make it virtually effortless for the customer to buy all of the products on the list at the same time. These alliances are an essential competitive weapon: On the Web, customers are likely to go to the service that provides the most convenience. Why force the customer to interact with someone else in an area related to your product (essentially letting another "camel's nose under your tent")?

Auto-By-Tel, originally designed to simply sell cars, offers an excellent example of product extension for one-stop shopping. Typically, there are several aspects of the process that are important to buyers:

1. Best price on a specific vehicle with specific features
2. The opportunity to weigh the advantages of buying against those of leasing
3. Financing
4. A hassle-free insurance experience

The company's fundamental purpose was to satisfy number one by affiliating itself with dealers who would offer the desired car at the best price. From there, Auto-By-Tel has evolved so that something as complex as buying a new car can truly be one-stop shopping, since a customer can get comparative prices while shopping and then work out financing and insurance before the ink on the purchase is wet.

For example, to help customers solve their insurance problems, Auto-By-Tel chose to partner with an existing provider of car insurance: AIG. AIG is prominently featured on Auto-By-Tel's Web site as a valued service provider. Auto-By-Tel even offers access to AIG's Bank Rate Monitor so that customers can be certain that these loans are competitively priced.

This partnership is just one more weapon in the company's fight to remain an industry leader.

ESTABLISHING STRATEGIC ALLIANCES
TO IMPROVE YOUR BOTTOM LINE

Strategic alliances are easy and possible through the Internet, and these selling relationships, arranged to be beneficial to all parties, can take companies a great distance in providing a "total solution."

Like Auto-By-Tel, AOL also has recognized the need to provide end-to-end service in various categories, one of which is its Real Estate Center, where it has linked with companies such as the Century 21 System, Intuit, and Countrywide Home Loans. In order to attract real estate business in a hypercompetitive world, AOL knew it needed to provide the complete solution for customers. It derives income by charging rent to those it invites on board.

"There is a shift taking place in cyberspace. Consumers have more access to fundamental information and that means power in purchasing and finance," says Rob Shenk, who oversees the area for AOL. "Consumers are now able to find the best real estate deals in an incredibly convenient new way. We don't believe these helpful resources can be found in one easy package like this anywhere else."

The center is comprised of five separate categories to simplify the process for AOL members: *Looking* (they can pull up and consider listings based on their specific family needs), *Mortgages* (in a partnership with QuickenMortgage, the ordinarily laborious process of researching and prequalifying mortgage loans becomes one of the easier parts of moving); *Selling* (members receive advice on selling their homes as well as new tax law information as it affects home sales; they also gain access to AOL classified home listings), *Renting* (through AOL's partnership with Apartments.com, the site lists over 5 million rental units; SmartCalc and Intuit calculators help members compare the benefits of owning with those of renting); *Moving* (the site also offers advice on moving, information about neighborhood schools, and access to other local information).

And Compaq Computer Corporation has linked with what the company refers to as best-in-class partners to create Compaq Entrepreneur Solutions featuring Compaq Online Solutions. Through partnerships with companies ranging from Microsoft and GTE Networking to e-Parcel and E-Stamp, the company provides entrepreneurs with faster and better ways to address the computing needs of small and medium businesses.

Although UPS's online initiatives are covered extensively later in the chapter, no discussion of strategic alliances should take place without noting that UPS has created an enormous number of partnerships as a way to better serve its customers and grow its business.

The financial arrangements associated with these strategic alliances are unique to each situation. However, they typically create a source of additional revenues for the host service and involve two elements:

1. The "host" seeking to provide the "total solution" seeks out the high-quality providers of the services or products needed.
2. The host looks for a top quality provider who makes the best offer. This typically involves the service or product provider's paying a flat annual fee or paying a percentage of revenues generated from the alliance (or both) in return for the business generated at the site.

BUILDING THE COMPANY-
CUSTOMER RELATIONSHIP

The Web makes it possible for customers to be treated like "one of the family." As an added enticement, companies are opening portions of their intranets to certain customers. Depending on the nature of the business, this can permit a customer to check on the status of an order, get directly in touch with technical experts, or check billing. It's a perfect way of deepening the relationship with your best clients and is being used by businesses large and small. Wickes Lumber Company in Vernon Hills, Illinois, has a system that permits registered customers to go online to check account information and order status.

Under partnerships between two companies, embedded extranets, where business partners can have retrieval of information from each other's sites, are another technological change that will provide a "total solution" for customers. The linked management information systems (MIS) will permit a rapid exchange of information between suppliers, resellers, and service providers. For example, one business Web site can query a second company's database, and results are delivered without having to leave the first Web site. The link is a strategic advantage because it eliminates a step for the user. Rather than going onto FedEx's Web site to find out when a delivery will arrive, a customer can query the FedEx database from the partner's site.

LEADING THE CUSTOMER
INTO UNCHARTED TERRITORY

As the Web redefines business practices, it creates opportunities for companies to add new aspects to core businesses. Typically, innovative products require a strong service component so that the buyer is positively inclined to use the new capability.

An in-depth look at the UPS Web site in mid-1998 shows one

example of a company working to build a new market—and at the same time realizing that it must provide the "total solution."

First, the company has added a number of important new features to its site. Some are directed at encouraging companies of all sizes to get involved in e-commerce: The site includes a lengthy list of companies that will work with customers and UPS to bring a customer's business online. While UPS's motive is to gain package delivery service when a product is bought online, it realizes that in order to build this market, it may benefit by providing extensive guidance for companies on how and why they should enter the electronic commerce arena.

Second, UPS clearly realizes that the Internet offers both a threat and an opportunity. Faced with the fact that fax machines and e-mail were eroding its document delivery business, UPS has defined its mission broadly and created the UPS Document Exchange to provide a secure way to send and receive critical business documents and other information—from video to audio to text—via the Internet.

In introducing these features (one a marketing tool and the other a new service), UPS is effectively saying, "Let us lead you toward the future, and provide the solutions you need." This is an intriguing proposition for most companies.

Fruit of the Loom has used a similar technique: It provides software and consulting to wholesalers to set up online storefronts. The wholesalers' customers, the screen printers, then search the wholesalers' inventory for T-shirts from a manufacturer. Fruit of the Loom only requires that its products be the first option offered to a Web buyer if another company's product is out of stock. Eight percent of Fruit of the Loom business is over the Net and its market share of online sales is more than 50 percent. In the real world, Fruit of the Loom and Hanes are tied at 40 percent of market share in the $5 billion industry, so to Fruit of the Loom, guarding its online lead is of paramount importance.

THE WEB CAN CREATE
NEW TYPES OF SOLUTIONS

In many cases, the Web makes it possible for companies to create new, formerly impossible, solutions that help to build their existing businesses.

Fortune 500 company National Semiconductor Corporation produces system-on-a-chip silicon solutions for the information highway. "Our focus on using the Internet as a customer support tool gives our customers an advantage over their competition," says Phil Gibson, director of interactive marketing for National Semiconductor.

With sales of approximately $2.5 billion and 13,000 employees worldwide, the company has created an extranet application that extends support to both its distributors and its direct sales force domestically and internationally. (The company reports that the savings to the distributors by ordering online amount to $20 million this year.) Those with access to the extranet have at their fingertips customized technical information and sales materials as well as an opportunity to give direct feedback.

By streamlining its support to direct and indirect sales forces, National is giving them back precious time to concentrate on other value-added activities with their customers. Furthermore, the feedback loop created between the company and the sales forces allows National to better forecast upcoming procurement and manufacturing challenges. This is just one of a growing number of Web investments that allow National to improve its customers' ability to make faster decisions.

Behavioral Healthcare Network (BHN) is an intranet for mental-health and drug counselors at cash-starved nonprofit agencies throughout the rural panhandle of Texas. The intranet employs state-of-the-art software that helps counselors evaluate patients' problems and makes it easy for health care providers at different agencies to track a patient's progress, thereby making it easier for counselors to coordinate care.

In this case, the online network creates a human network. When a counselor sees a new patient, he or she logs on to BHN's Lotus Development Corporation Domino server, asks questions, and fills in the online form provided by BHN's assessment software. The information is recorded and tracked on the intranet during the patient's treatment. If inpatient treatment is necessary, counselors can fill out a Web-based form and get the person on a waiting list for the nearest facility, which happens to be in Amarillo. Later counselors can check detailed logs of what happens to the patient in the hospital. While improved patient care is an immediate benefit, the system also positions the organization for the future.

THE "TOTAL SOLUTION" MAY BE MORE DEMANDING ON THE NET

The evolving world of Net shopping is almost certain to create new demands as to what constitutes a "total solution." For example, comparison shopping is now so ingrained in the Web buying experience that the "total solution" may ultimately mean displaying, on a regular basis, the prices or products offered by your competitors. This strategy will be fully explored in Strategy 7.

THE BATTLE PLAN

As I discussed at the start of this chapter, the goal in providing the "total solution" is "to be so good that your customer doesn't need to look anywhere else." In a hypercompetitive world, this is no small task.

I believe the key to providing the "total solution" on the Internet is much the same as it is in the brick-and-mortar world—you have to listen to your customers, a process made easier via the Internet.

I suggest asking as many customers as possible three questions:

1. What else could we be doing for you that would be valuable?
2. Is there any aspect of our services that needs improvement?
3. What are your main business problems today?

There is an important reason for asking the last question. In many cases, you may be able to provide your customers with new services or capabilities they don't yet know exist. One of the goals of this new hypercompetitive world needs to be making the creative jump from the problems customers describe to products that meet these needs.

Companies that find online ways to satisfy customer needs will find they can thrive in this era of hypercompetition. Those that simply try to move their "real world" business to the Web are more likely to be disappointed in the resulting sales.

Market Relentlessly

When you mention the Web and marketing, most people think, "Here comes a discussion of banner ads." I don't want to minimize the importance of "cracking the code" with regard to bringing visitors to your site. Over the next several years, companies will literally spend billions of dollars on this task. However, this is valuable only if you have first managed to create a compelling presence on the Web, one that survives the "First-Time Visitor Home Page Test." (If first-time visitors to your Web site never go past your home page, it's a pretty good bet that you should investigate a more compelling approach.)

Success will first depend on mastering the type of marketing involved in engaging visitors at your site.

There are as many ways to market online as there are online vendors. While optimum use of the Web will differ from company to company and industry to industry, companies must commit themselves now to mastering the revenue-generating potential of this new medium to the fullest extent possible. Consider the Web initiatives established by these businesses:

In the chemical industry, it can take hours to check through as many as two dozen print catalogs to compare price and characteristics of chemicals used in research. Sometimes the catalogs are out-of-date; often the one item needed can't be found. That's why Chemdex was founded—to be an independent market manager, bringing together buyer and seller and streamlining the sales process. Drawing from small and large vendors alike, Chemdex is getting its data in front of the nation's estimated 300,000 researchers. The alternative is a print catalog, glacier-like in its updating process when compared with the Web and much more cumbersome to use.

Midwestock, a midsize licenser of stock photography in Kansas City, Missouri, a company that formerly could not afford to publish a catalog regularly, can now market one to one via the Web. To make its selection manageable, Midwestock publishes only 150 of its 150,000 images at a time, rotating the selection regularly. However, if a customer calls to request a photo of a grizzly bear, for example, the company can post a selection of its best bear photos on the site, with the Web address needed for access.

While many of the ideas and examples that fill this book are oriented toward generating revenues and therefore fall within the broad definition of marketing, this chapter will concentrate on establishing the best atmosphere from which to sell. This chapter is not a "how-to" discussion; successful Web marketing is the subject of entire books dedicated to this subject, including my own *NetMarketing*. Within the context of HyperWars, it's more important now to focus on how you should think creatively when it comes to marketing on the Web.

DEVELOPING A MARKETING MIND-SET: EXPERIMENT, LEARN, EXPERIMENT, LEARN

The world is now full of Internet initiatives that may have started as experiments but have become central to the lifeblood of the company. As a matter of fact, the idea of an intranet was actually nothing more than a successful experiment. When Web browsers were first developed, no one was focused on internal corporate use. Later on some unsung hero of the information age realized that the Web and browsers, combined with security features, could achieve one of the holy grails of corporate computing: the ability to link employees using different types of computer platforms at many different locations in a single secure network, which could then be packed with valuable information for company employees.

There is a consistency to successful companies' methodology: They have pursued a process of trial and error followed by incremental improvements. The medium is so new, and customer response so unpredictable, that it is impossible to know what will be the most effective mechanism for using the Web before you actually try it.

Within this framework, three elements are central to sales success:

First, the company must be willing to commit the resources — including money, talent, and executive-level interest — necessary to making the Web a valuable part of the revenue mix.

Second, there needs to be an overriding interest in experimentation. Companies recognize from the start (and budget accordingly) that they will want to try many different approaches to engaging potential customers online. This leads to the following practical necessities:

- Companies must establish a way to evaluate Web success through some form of measurement. Those creating the Web initiative should identify beforehand what's important so that it can be measured. This gauge could be traffic (number of visitors), it could be sales, it could be unique prospects who sign up for an e-mail mailing list, or it could be a far more subtle indicator.
- The company plan should include a number of different alternatives to test over time. From the outset, there should be a spirit of continual testing and evaluation. Sometimes companies fall prey to the natural tendency to never fully move beyond what was started. I strongly urge you not to

let inertia—or varying degrees of success—prevent further experimentation.

Third, companies must respond with flexibility to what they learn. If you discover something with great potential, give yourself the freedom to take it to the next step quickly instead of remaining wedded to the original plan. In a HyperWars environment, quickly moving forward can be quite a valuable competitive advantage.

CUSTOMER SERVICE AS PART OF YOUR PLAN

Companies that succeed in integrating the Web into their revenue efforts recognize that the overall customer experience is of tremendous importance. Creating excellence in customer service via the Web is integral to a strong sales and marketing effort.

Pre-Web, consumers who needed follow-up information on products were obliged to read instruction booklets, call toll-free numbers, or phone a local store staffed with authorized dealers or representatives. Now, all this can be centralized at a single Web site, and both front-end and back-end customer service can be provided at a lower cost.

This is a critical battleground in HyperWars. In industries where products and prices are similar, what makes the difference between success and failure is customer loyalty, and what often builds customer loyalty is attentive and intelligent customer service.

While Strategy 6 provides many different examples of ways that companies are creating a "total solution" for their customers, one thing you should keep in mind is that the technology is going to make excellent customer service better and cheaper over time. Broderbund Software, Inc., is one of many companies that have found they could actually reduce their service-representative head count because a good Web site actually lets customers answer their own questions. As early as 1997, the company estimates, its site handled at virtually no cost volume that would have cost $265,000 to manage in the call center. Broderbund saved money while sending away satisfied customers.

INCREASING SALES BY KEEPING CUSTOMERS UP-TO-DATE

Could you sell more product if you had an inexpensive way of keeping customers up-to-date about new developments? Because the Internet permits instant publication of news and product updates, merchandise listings and availability can be updated at any time.

Early-alert software has helped companies notify individuals of upcoming events in a way that will serve as a marketing model for e-commerce.

The low cost of e-mail combined with the convenience and ease of response make it a strong vehicle for local businesses to use for increasing their sales. For example, a local pest control business might send an e-mail message to all of its past clients asking if they would like to schedule a visit for annual maintenance. A yes reply to the e-mail by e-mail (How easy is that for the customer? All he or she needs to do is click on "reply," type "yes," and hit "send") would result in a call from the company to schedule an appointment. By serving as a reminder service, the local pest control business can maintain an ongoing relationship with its clients, and clients are happy because it eliminates their need to remember to schedule a visit. Similarly, I know of a local veterinarian who uses e-mail to remind customers that it's time to bring their pets in for a vaccine.

Retailers can also benefit by establishing e-mail communications that keep customers up-to-date: Lands' End provides an e-mail service that alerts customers to special offers. The company is one of many retailers that offer online-only sales on its merchandise.

MARKETING OUTREACH

One of the benefits of the Web is that companies find that they can reach more than their original target audience. The Web is filled with examples of companies that are reaching entirely different markets than those they serve in the physical world.

West Group, a legal publisher based in Eagan, Minnesota, is using its Web site as a third sales channel, along with a field sales force and direct-mail/telemarketing efforts. It has discovered a receptive new customer base among the general public, particularly for sales of what it calls its "nutshell" line of books, books that offer general descriptions of important legal concepts. Not only is West Group selling to a market that extends beyond lawyers and students, but it is finding that its profit margin on the Net is far better, since the cost of selling a $20 or $30 item is lower on the Web.

Before the Web, many companies lacked an effective means of selling their products and services to smaller businesses or individuals and had no cost-effective mechanism for providing these groups with appropriate customer support. As Web sites have developed into extremely efficient means of taking orders and providing support, a number of companies have been able to go after these groups as new markets. One example of this effect is the S&P service for individuals discussed earlier. Prior to the growth of the Web, S&P did not have a viable means for selling and distributing a product to individuals.

BUILDING COMMUNITY

At its best, the idea of a community is to create a place where product enthusiasts can gather and discuss aspects of the product they enjoy. This type of forum gives enthusiasts an even stronger feeling that they are connected to the product, and that they are not alone in their interest in the product, and it may spur sales of high-margin peripheral items related to the product.

The Internet, with its myriad chat capabilities, can certainly encourage community bonding. Take a look at Gund's site for teddy bear buffs; Reebok's site where sports and fitness enthusiasts gather to chat with each other, check in for fitness tips, and browse for information on upcoming events.

CHECKING THE COMPETITION

Some time ago, *Entrepreneur* magazine ran a cover article titled "The 101 Best Marketing Ideas Ever." The magazine asked me to write the section of this article addressing online marketing. One of my rules was essentially "Check your competition's site at least once a week."

A careful examination of a company's Web site can lead to important insights about the products the company is planning and how the company is positioning itself to compete. No one has a lock on good ideas. If the competition is doing something terrific that is appropriate for your product as well, my advice is to match it as quickly as possible.

And obviously, if you can benefit from its Web site, it may be benefiting from yours. I have known companies that have hesitated to use the Web as a marketing tool because they didn't want to tip off their competitors as to what they were doing. Don't let this bother you. My advice is always "Full steam ahead." In this hypercompetitive age, it's better to create a great site and reap the benefits.

PRICING AND MARKETING STRATEGY

In all businesses, pricing is a difficult, often complex decision: Should I price low to build market share and hope to make profits through volume? Or should I price high and make a good profit on each sale of my premium product?

On the Web, these decisions have one additional element that can also add to the complexity: whether a piece of the product should be *free* in order to build consumer interest. (The issue of strategically using free products is the subject of Strategy 8.)

As you consider pricing products in a HyperWars environment, it is important to keep several guiding principles in mind:

First, your product or service may not need to be the absolute lowest price, but you do need to be competitive. The reason I so strongly urge you to cut costs in Strategy 3 is to give you the latitude to be among the low-cost producers who can effectively compete on price.

Second, if you're weighing whether to drop your price to build market share or to keep your profit margin high, opt for building market share. As I have discussed, the Web accelerates the natural tendency of any industry to move toward consolidation. A product, service, or "store" is anointed as best of breed, word spreads quickly, and this "winner" quickly becomes the market leader. It's far easier to build a profitable business when you have a large market share—even if the margins required to obtain this position were slim—than it is to try to move customers from an existing relationship.

As discussed throughout, once you establish a relationship with a customer, there are any number of ways you can work to build loyalty, or "lock in" that buyer. The company trying to convert a customer is fighting all of these mechanisms as well as the natural inertia of a happy customer to stick with a supplier with whom he or she is comfortable.

And third, recognize that (unfortunately) it's a harder world for premium products. It used to be that if you launched the premium product in a category, you could look forward to premium returns based on the higher prices you could command, for quite a long time. Now, faster development cycles, free offers that may not be nearly as feature rich but provide the core value of your product, Web-based comparison shopping, and bundling (discussed below) have narrowed this window.

Build in Comparison Shopping

More and more software comparison-shopping agents have come into use, and at sites such as Yahoo! and Excite!, consumers can request simultaneous shopping comparisons from multiple retailers about price as well as features.

Polymerland, a GE subsidiary, lets engineers search its online database of polymers to find the resin that has the exact quality needed. When the site added a shopping bot, traffic increased eightfold, with one customer in four using the site regularly.

This type of strategy encourages visitors to start out at your site, knowing that you'll give them the information they need, and that you are likely to get the sale by "default" or inertia.

As discussed earlier, the online computer hardware and software store New England Circuit Sales (NECX) employs a similar strategy. The online store launched a capability that lets users compare its prices with those of half a dozen competing online retailers, even when its own price is higher. The result? NECX's sales ($5 million a month) increased 20 percent in the first three weeks. This type of strategy is clearly effective.

The Problem of Bundling

The entire cycle of product innovation and development has accelerated for all types of companies. As a result, competitors are quicker to match new innovations, and this brings with it the threat of bundling.

The term "bundling" refers to a situation where good individual products are combined with other products to create one unit for sale. For example, spellcheck was once sold as separate software; now customers would truly be shocked to buy a computer that didn't have spellcheck "bundled" in as part of its word-processing package. As competition grows more intense, bundling becomes a bigger threat. Consider this scenario:

- A company creates a superior product or service and "wins" in the marketplace.
- A different company comes along and says: "We want to include your product as part of our larger one and we will pay you a bulk rate."
- If the company with the number one product refuses ("Why should we essentially give away the dominant product in this industry?"), the company with the proposed product bundle then approaches the second-tier competitor. The competitor's product may only be 80 percent as good, but the competitor decides to accept the bundling proposal.
- Now the competitive dynamics are entirely different. If the product bundle is a success, then potential customers already have their product—albeit only 80 percent as good. The choice now presented to consumers who purchased the product bundle is whether or not it's worth it to spend more money and buy the superior product alone for an additional 20 percent of functionality.

You may have a far superior product, but if the customer believes he or she already has a product that performs similar functions 80 percent as well, it's challenging to convince him or her of the need for your product as well.

My advice: If a bundling opportunity arises, consider the extent to which it will be a threat. Then, either participate or create your own competitive bundle with other partners, which will also keep your product at the cutting edge of competition.

SAVVY MANAGEMENT
OF EXCESS PRODUCT

Nowhere but on the Web can companies so effectively manage their inventory. If a product is selling poorly, a company can move it to the front page of its Web site and tout it as a "special offer."

Online auctions via specialized auction sites and other services provided by new digital age middlemen have also proven to be a superior way to manage excess inventory. And Fastparts, Inc., a site where companies around the world buy and sell surplus electronic parts, may one day all but eliminate the need for traditional parts brokers.

TARGET MARKETING

Most companies are still in the very early stages of exploring the potential of target marketing or selling one to one. Nonetheless, it may ultimately prove to be a powerful weapon in a HyperWars environment.

As discussed in Strategies 5 and 6, a unique form of selling is evolving on the Web. Consumer product sites will recommend products based on what individuals have purchased before, while high-tech companies may automatically provide their business-to-business customers with updated products.

Smart companies will build this way of thinking into their marketing plans. They may ask consumers which of the products they have particularly liked (or not) and why, so that they can provide increasingly more relevant suggestions. An online "auction house," First Auction, uses a special marketing software to profile customers in order to better judge what it should be selling.

Still other companies such as Intel and Met Life are asking site visitors—who are prospects—what they'd like to hear about. Potential customers sign up because they want to hear from the company—there's no better mind-set for converting a prospect into a sale. Seth Godin, the founder of YoYodyne, an interactive marketing agency, calls this "permission marketing": The customer has told you that he or she wants to receive information about products and services via e-mail.

Some variations on this idea include Security National Mortgage Corporation (which allows prospects to sign up for a service that notifies them by e-mail when specific mortgage rates become available) and the Electronic Newsstand (which sells print magazines online and sends a regular e-mail discussing the content of some of its offerings).

Similarly, companies are using the idea of permission marketing after visitors have made a purchase. When a purchase is made online, most sites will ask if you want to hear about new offerings and other product news. In this way, firms are able to establish an ongoing communications channel, and product loyalty, with their best customers—people who want to buy more of their product.

An important aspect of well-implemented Internet marketing is that you are contacting customers and prospects because they have indicated they want to hear from you. Toward this end, e-mail is most effective when used sparingly. Companies should resist the urge to spam (send unsolicited messages to individuals with whom you have no relationship). It's annoying to recipients and will actually dilute or destroy any company goodwill you might build through sending helpful news to customers and prospects.

The potential for target marketing and building ever-deeper customer relationships offered by the Web is enormous. You don't, however, want to "upset" customers by suggesting you have in any way violated their privacy or collected information they did not realize you were storing. As a result, the key to success will rest with companies that strike the right balance—between customer permission, privacy, and customization. Issues related to privacy were discussed at length in strategy 5.

MARKET RESEARCH MADE EASY (AND AFFORDABLE)

Market research over the Net is faster, reduces costs, and in many instances, generates better results than traditional methods.

When a company considers conducting market research, three central questions should always be examined: How expensive will it be? How fast can we do it? Will we derive valuable results that will be worth the high expense? The typical means of conducting this research range from expensive options such as massive telephone calls and focus groups to direct mail, which is both costly and slow in yielding results.

Companies are finding that they can accomplish excellent results with virtual focus groups at a far lower cost. One research company reports a 50 percent response rate to an e-mail it sent soliciting consumer input on four print-ad concepts for a client. As compared with responses that are often in the single digits, this is an extraordinarily high response rate, which led to valuable information for the client faster and at a lower cost than conventional means. In a separate effort, Avon found that an online survey more accurately predicted the demand for a product than did traditional focus groups.

This type of success is not an accident. As a medium, the Internet is well suited to companies that want to conduct market research in a smart, nonintrusive manner. Virtual focus groups are nonintrusive because the consumer can participate when convenient (as opposed to being phoned during the dinner hour), they are visual (consumers don't have to go anywhere to register votes on appearance), and they take only as long as the consumer wants to spend (no one is dependent upon a human survey-taker to go through the questions).

Here's how online research might be conducted: A typical survey for JCPenney might involve reviewing sixty swimsuits online, with women clicking on their likes and dislikes. JCPenney estimates that it pays approximately two-thirds less for an Internet customer survey.

The subjects for online research can be recruited through a number of means, depending on the specific goals of the research. In some cases, it may make sense to e-mail current customers and ask them to visit a special area in a Web site designed for conducting the research. Or a company might randomly "stop" visitors to its site through advertising banners or more intrusive pop-up messages, inviting these prospects to visit, through a simple click on a hyperlink, the research area of the site. Unlike in-person focus groups, customers and visitors will typically participate in this research without any special incentive—the inconvenience is limited, and I suspect participants find it interesting enough that they don't feel a need to be compensated.

Winners in the future will be companies that can change course quickly, keep their costs down, and understand their consumers very well. All of this suggests that in any business where market research is necessary, studies conducted faster and at lower cost through the Internet should be seriously considered. Moreover, smaller companies, which might have previously considered such efforts beyond their means, should now be taking a hard look at the possible benefits of conducting market research.

VISIT-MONITORING

A Web site is in itself a source of rich market research information. By looking at what areas customers and prospects most visit at a site, companies can learn a great deal.

Today sophisticated companies monitor customer call centers for the types of calls they are getting. This, in part, helps to guide them in where they may have product areas that need improvement. The Internet creates a far more powerful capability. Because Web sites can be set up to track not only how many people visit but where they are coming from, companies are finding the Net instructional for better identifying customer needs as well as targeting their markets. (Note: These initiatives do not involve tracking specific individuals, so they do not raise privacy issues.)

Two very different examples show the range of information and usefulness such information can provide: One company that uses this tracking to determine specific areas of opportunity is National Semiconductor. This large chip maker monitors activity at its external Web site for indicators of potential demand for products, and in one case this monitoring led to changes in a product that helped to increase sales by an extraordinary amount: from $20 million to over $100 million.

A very different example involves car manufacturing. If businesspeople could prophesy what customers will buy, they'd almost always succeed, and those who are listening via the Internet are enhancing their opportunity for just that. As a result of activity at its Web site, at least one major car manufacturer has increased sales by changing the proportion of cars manufactured in specific colors in order to increase sales.

THE BATTLE PLAN

In an effort to determine a marketing mind-set that is right for your company, you need to consider at least three distinct areas:

1. The allocation of resources, including valuable people, to create this effort
2. How the Web can be used to creatively market and sell a company's existing products or services
3. How the blending of the Web's capabilities with a company's core expertise may lead to the creation of new sources of revenues

Certain elements of this kind of marketing strategy can lead to success, or doom the effort from the start

First and foremost, top leadership must be supportive of the initiative. If a company's leader doesn't stand firmly behind the effort, then talented people will shy away from participating and budgets will inevitably be too slim.

Second, the group leading this initiative must adopt a wide variety of different initiatives in order to assess what works best. As noted earlier, one of the Web's great capabilities is the ease with which an idea can be tested with real prospects who are visiting a Web site and then just as easily be modified or withdrawn for something else. It is critical that in these early days of online commerce, companies adopt this approach.

Third, this multiple initiative approach needs to mesh with the strategy discussed earlier—that speed is everything. In a word, successful competitors on the Web will be fast-moving and have real "guts." A company can spend seven days, seven weeks, or seven months, on a single test. Decision-makers must be willing to make a creative leap: In today's environment, you don't have time to know everything that you would like. It's more valuable to start new marketing initiatives and refine them as you go than to try to know everything upfront.

By testing multiple approaches and using the Internet to gather information, companies will gradually become more sophisticated in their marketing efforts and therefore more successful in the evolving hypercompetitive environment.

The Magic in "Free"

I spent a large part of my early career at Time Inc., one of the nation's most successful direct marketers of subscriptions for its extended family of magazines, Book-of-the-Month Club and Time-Life Books, and the guiding mantra throughout the hallways of the Time-Life Building has long been "The most powerful word in the English language is 'free.'"

Today messages such as "*free* trial," "first thirty days *free*," and "*with no obligation to buy*" are part of the standard business tactics used in trying to build interest among potential buyers. So what makes "free" special—and necessary, when it comes to the Internet?

The Internet creates a very different sales environment because of the ease with which shoppers can move from site to site. In a world where detailed price and feature comparisons can be conducted in under sixty seconds, good pricing is important, but something extra is often required to turn a prospect into a paying customer. By offering something for free, companies can establish a "relationship" with potential customers so that when they are ready to make a purchase, they are inclined to buy from the company they "know."

A well-executed strategy for managing "free" can lead to the successful recruitment of new, profitable prospects and a growing business. A poorly conceived one will cost you in both time and money. This chapter explores how companies are using "free" on the Internet to build awareness and provide value using various methods, including a new form of "try before you buy" unique to the Internet.

THE SPECIAL NATURE OF "FREE" ON THE INTERNET

Those who are launching Internet businesses realize that in the world of HyperWars, you can't afford to wait for a customer to come to you—you need to be out there, fast and first. In some cases, companies are finding that they can't start to charge for products until they have demonstrated their ongoing value to consumers by offering "something for nothing."

There are four factors that have combined to increase the value of using "free" as a marketing tool on the Internet:

1. The "real-time" communication aspect of the Web makes it possible for companies to provide consumers with instantaneous value.

Visitors to the Web site of Acumin, the vitamin manufacturer discussed earlier, can take a simple health and lifestyle quiz and instantaneously receive recommendations as to what vitamins and nutrients they need to take in order to get on the path "to a healthier lifestyle and complete nutrition." Of course, visitors can purchase the vitamins direct from the Web site or by using a toll-free number.

Without paying a cent, gardeners can turn to Garden Escape's Web site for planning their gardens. The company's Web site permits registered visitors to file away their plans to work during repeat visits. Items the gardeners may eventually want or need are available through the site.

2. The technology makes it possible (and inexpensive) to offer costless incentives that provide real value.

At a Web site, a company may analyze visitors' insurance needs or determine what sort of makeup is right for a particular visitor's skin type. For most companies, this is a fixed cost—the company pays once to have the software created. Then, since there are no 800-number charges on the Internet, there are minimal incremental costs involved in allowing an unlimited number of potential customers access to the information. Just as a company budgets how much it will spend on radio advertising, now marketers can establish a dollar amount the company is willing to spend on heightening awareness through offering a free service over the Internet. The difference? Radio bills are recurring; dollars for Internet software are spent once.

3. The absence of communication costs (and sometimes distribution costs as well) between the prospective customer and the product provider makes it affordable for companies to make their free offers continuously available.

With no long-distance or postage charges for either of the communicating parties (the potential buyer and the potential seller), Internet companies, ranging from those distributing free information to music, are finding it reasonable to offer something for nothing on an ongoing basis.

4. The low incremental cost of providing many products or services once they are developed for the Internet means that companies may use a trial offer of the service as a "cyber loss leader."

Like the retailer who advertises a low-priced item in order to entice customers into stopping in and also picking up some more profitable items, the Internet's "free" brings people in and then most sites hope to extend the sale.

THE BEST THINGS IN LIFE ARE FREE
(ON THE INTERNET)

Like a carnival midway, the Internet offers attraction after attraction
—a visitor could be happily occupied for hours. But the Internet
provider and the carnival hawker face similar challenges when it
comes to fighting for attention. On first glance, a visitor doesn't necessarily see a compelling reason to visit one site (booth) or another.
That's where the "big teddy bears" come in. In a life-and-death
struggle to compete for attention, companies are using various
forms of "free" as a way of indicating that customers at their site will
bring home the "biggest bear."

Using the Internet to Build Awareness
and Demonstrate Value

Both online and in the real world, "free" accomplishes several
important goals that have to do with building awareness:

- It attracts interest. ("You want to give me something?")
- It introduces a new product to potential customers.
- Familiarity with the product helps create desire.

When a product is offered free to build awareness, the intention
is that the consumer will become familiar with the company and
its products and decide that it makes sense to purchase other products from this company. Here are some innovative ways companies have approached this strategy:

After death and divorce, moving is the third most stressful of
life's events, and in order to make it less so, Atlas Van Lines, a $400
million mover of household and high-value goods, hopes that its
Web services will make the relocation process less harrowing. Customers can get everything from information on estimates to tips on
packing. Atlas offers a moving schedule to help customers plan for
their ordeals. Visitors enter their moving dates and get back a per-

sonalized schedule, complete with lists and reminders of what to do four weeks before the move and so forth, to print out and hang on the refrigerator. Site visitors can request specific reminders by e-mail if they so desire. Here, Atlas has positioned itself as a friendly expert who knows how this should be done. When it comes time for the customer to actually select a mover, it would then be natural for the customer to consider Atlas.

The book business is also making good use of "free" by offering lots of information in order to build interest in products and keep people at their sites. All the online booksellers offer reviews and analyses of many of the books, and these write-ups can range from a reprint of the publisher's press release and comments by the author to reviews posted by customers/readers. Barnesandnoble.com offers real-time chats with authors who have included Tom Clancy, Anne Rice, and Kareem Abdul-Jabbar, and more than 2,500 people showed up for the Drew Carey chat—this is an audience size authors who do in-store signing on book tours would die for.

On the Internet, the low-cost possibilities for attracting attention are high and go well beyond the physical supermarket equivalent of handing out small samples to taste new products. As might be expected, the best implementations are creative and customized to the specific business or product involved.

UNIQUE TO THE WEB: TRY-BEFORE-YOU-BUY SOFTWARE

The Web creates yet another unique capability: Companies can make full-scale products available to consumers for a limited period of time. This is particularly appropriate with software, and it is applicable with popular titles such as personal finance software as well as sophisticated business-to-business applications.

In the past, software companies frequently made "demo" versions of new products available to prospective customers. These efforts, however, faced certain inherent limitations. First, the software itself was not the full version of the true product—so the prospective customer's ability to evaluate the product may have been somewhat limited. Second, software companies had to strategize and spend money getting disks with these demo versions into the hands of prospective buyers.

Now, what is known as "trialware"—complete versions of software you can "test-drive" or "take for a spin"—has emerged and can be downloaded directly from a software retailer's site, from the manufacturer's Web site, or through links from other high-traffic sites frequented by customers. Trialware is superior to demo versions of software: Demos are made with limited capabilities, while trialware offers a user the opportunity to experiment with the full capability of the software. Typically, the software carries a "time lock" of some kind, rendering it useless after a trial period, such as thirty days, unless the consumer contacts the company, makes a purchase of that software, and is then given the code to prevent the time lock from taking effect.

So what is a company's typical success rate at making a sale? Trialware offers one hook that software publishers rely on. Test-driving software is better than test-driving an automobile. When you test-drive a car, you usually don't put all your belongings in the trunk and fill the glove compartment as well. With software, users often enter lots of critical information, and when the clock runs out on their trial, they are left with a big incentive to buy the application—the cost in time and effort to reenter the data in a different program.

In the long run, this type of purchase usually leads to lower costs for consumers. Since software companies don't have to incur brick-and-mortar retail distribution costs, they frequently pass some of these savings on to consumers. Thus, trialware is generally less expensive to purchase than shrink-wrapped software.

There is little question that software publishers and distributors will rely more and more on try-before-you-buy technology as the market for electronically distributed software grows. According to Forrester Research, 50 percent of all software will be distributed electronically by the year 2000.

One example of how this trial strategy was particularly effective in the software arena involves Network Associates (formerly McAfee Associates), a leading maker of virus-protection software based in Santa Clara, California, that attributes its success to the power of "free" and the new, associated economies. The company has found that try-before-you-buy has also helped extend the company's reach in other ways. By granting thirty-day "evaluations" on its products, the company has more people reporting viruses, and as a result, Network's software has identified an increasing number of viruses. The company now has 25 million users, annual revenues of more than $200 million, and a market cap of nearly $3 billion. What's more, there's more value for the price: As its market share grows, so, too, does the feedback from customers about viruses, thereby delivering to the company information that further enhances the product. As a result, users are willing to pay for updates more frequently.

This is only the beginning of the innovations that may occur as a result of the widespread growth of the Internet. At this writing, there is indication that "rented software" may become popular. In return for a monthly fee, software distributors or the manufacturers themselves will provide users with what is always the most current version of a piece of software.

This marketing approach applies to software of all types, popular titles such as personal finance software as well as sophisticated applications. And winning companies will not be afraid to experiment with a variety of approaches. For example, for April 1998 tax filings, Intuit made a version of its boxed Turbo Tax software available for use online (consumers didn't have to download or own it). The price for this new, online service was only $9.95, as compared with significantly higher prices to own the software.

Currently there is one primary limitation on try-before-you-buy, and that is bandwidth constraints. Slow connections to the Internet means that it takes more time for large pieces of software to download. Over time, as fast connections become ever more common, this constraint will disappear.

TRY-BEFORE-YOU-BUY EXTENDS BEYOND SOFTWARE

As personalized services are increasingly available over the Web, companies are also able to provide potential customers with real value on a trial basis while incurring only minimal costs for providing the free trial. One example of such a service is practical, the Online Weight Loss Clinic. Using expert technology, this Web-based service is able to provide potential customers with a free personal profile and list of foods that are appropriate to help a potential customer lose weight. In this way, the potential customer gets to experience the service and see if it's worth buying.

Here are other types of products that give consumers a taste of what is to come:

A customer who logs on to one of the sites selling music can listen to snatches of songs before deciding whether or not to buy the full CD. Technology is such that companies with certain software can issue "licenses" to play a certain CD on a personal computer, and this temporary copy could then expire after ten to twenty days.

Comprende, an online translation service, offers to send a free e-mail in another language for customers who want to try its service.

A quick stop at the Microsoft site and a click on "Free Stuff" will provide visitors with libraries of demos, add-ons, and beta versions for each of Microsoft's software products.

THE FREE FRONT PORCH

Everyone blames computers and television for lessening person-to-person communication—but this was before the Internet. Today people around the globe are benefiting from a plethora of online information being shared by companies and individuals alike.

What do you do if you're a financial publisher that sells hundreds of millions of dollars of industry analyses, performance data, and corporate earnings forecasts to commercial investment houses and brokers, and you want to expand your reach to the "little guy"? Why, you offer information for free. Or at least that's what Thomson Investor Network (TIN), an $8 billion-a-year publisher, is doing. To fight back against the flood of upstart Web sites offering free data, investors can use TIN to track stock prices of up to fifty companies in real time. Each quote comes with a TipSheet that includes a twelve-month stock chart, data comparing performances of similar stocks, and buy, hold, or sell recommendations from services that usually must be paid for. For an additional $9.95 a month, investors get in-depth company research, screening reports, analyses of insider trading, and e-mail alerts of news likely to impact stock performances. Many of these paid services are streamlined versions of products that sell for hundreds of dollars each year to professional money managers. Why do this? Exploiting its rich data to attract customers offers Thomson the opportunity to catch up with better-known Web sites like Quote.com and Microsoft Investor. Executives at TIN feel that the company will develop lots of new revenue streams. To begin, there's the $9.95 a month the company collects from individual investors, and advertising revenues are rising as companies such as Charles Schwab pay to sponsor the free TipSheets. Thomson is also negotiating to get a cut of trading fees, if brokerage houses and mutual fund providers come to see TIN as a natural place to acquire customers. While maintaining a high-traffic Web site is costly, Thomson's strategy acknowledges that there is an investment cost to pay for being a leader in the marketplace.

Information service providers, such as Thomson, are finding that offering part of their services for free is the right "bait." By attracting visitors, the service is able to generate ad revenue (although to date rarely enough to make the services profitable). More in-depth information—along with tools for manipulating this information—is then only available for a monthly paid subscription.

By offering certain information for free, a company hopes to develop a following of people who appreciate the value and quality of the information available at the site and want to know more—and who will eventually be willing to pay for it.

Some companies are just beginning to get their fee-for-information toes in the water. Companies such as *Business Week* are now charging for their online content that was formerly free. Others, like *The New York Times*, charge for use of their extensive archives while making the rest of the information available free. *Money* magazine operates a free area, and has also introduced Money.com Plus, a paid area with premium financial services information and analysis tools. While name-brand companies and companies that have already developed consumer loyalty are likely to succeed in charging a fee for service, new and less recognized companies will find they need to take measures to develop customer loyalty first.

USING "FREE" TO GENERATE SALES

Amid intense competition, companies are also using "free" to capture specific information in order to build later sales of related products. Any good salesman knows that the more you can get a customer to talk about himself, the more specific—and the stronger—the sales pitch can be. Interactive sites established by financial service companies have done an excellent job of getting people to "talk." By plugging in their own financial information, visitors can set goals, perform calculations, and learn more about various types of appropriate financial products. Again, it's worth noting that consumers are very concerned about online privacy, and companies need to be explicit with site visitors about what they will do with any personal information they collect.

Earlier, we discussed the evolution of digital age middlemen. In most cases, these middlemen put buyers and sellers together in an innovative way. One key to the success of these businesses is that they attract eventual buyers by providing them with free services. Then, because these buyers are a highly targeted audience, the service provider can find ways to offer them products it knows they will want. A cautionary note, however, is necessary here. You must have a plan for building revenues before creating your free community. Your plan may change, but at minimum you want to know that you are aligning all of your efforts in providing free value toward your goal of generating revenues.

Free e-mail reminder services that are offered by many Web sites operate similarly. By e-mail, these services remind people about important upcoming events, providing tremendous convenience. (In my house, if we miss sending a card to a second cousin on her anniversary, it's a tragedy—but I confess this date is not top of mind. So an e-mail reminder a week beforehand is of real value to me.) Typically, sites that operate reminder services also offer gifts for the occasion. This is an additional convenience for the consumer, so it represents a win-win situation.

As discussed earlier, eToys lets visitors register for a reminder service for holidays and birthdays. As a result of the reminder, eToys (or any "reminding" merchant) gets the "first crack" at the prospective customer's business for that occasion.

Another example of how a free product can feed a related business is the Internet browser. Both Netscape and Microsoft give away their browsers because attributes of the browser will, hopefully, lead users to ways these companies make money. In Netscape's case, its home page is among the most visited sites on the Web, because the browser is preset to take users there when they log on, and most users never change it. As a result, Netscape can use its site to sell third-party products and generate advertising revenues. While the specifics of Netscape's arrangements are confidential, my educated guess is that as a smart marketer, Netscape has structured its deals such that the more successful a third party is in selling products and services through Netscape's site, the higher Netscape's compensation.

But when it comes to generating sales, no marketer can forget about the good old-fashioned way: coupons. At allergy medicine Claritin's Web site, visitors can get tips on coping with allergies as well as coupons for Claritin products. At other sites, Web surfers will find coupons to save money on items as varied as pizza and office supplies. (A preliminary registration requirement keeps people from printing out coupon after coupon.)

USING BOTS TO BUILD BUSINESS

One good example of a bot that offers a free service for customers and, in turn, builds business for the company is RateTracker, a piece of agent software that is licensed to individual mortgage lenders. RateTracker monitors what's available from individual lenders and notifies earlier visitors (who have signed up for the service) by e-mail when specific types of mortgages they've requested become available. (Check Security National Mortgage Corporation's site to see RateTracker in operation.)

It's worth noting that this type of bot represents a highly profitable lead generator for companies offering mortgages. When consumers learn by e-mail that the type of mortgage they are seeking is available from a certain company, they are likely to buy from that company.

THE BATTLE PLAN

As competition intensifies and companies build on the idea that the first step in any sale is to develop a relationship with the potential customer, I anticipate that companies will use "free" even more competitively.

As a result, for businesses of all sizes, there is a need to master this evolving form of selling. Free resources—intelligent recommendation systems, e-mail newsletters, in-depth information, analysis tools, and bots with specific uses—should all be assessed against three very specific questions:

1. Is what I am providing prospects free today designed to start a process that ends in a sale?

2. Is there a way to measure whether this process is working?
3. What else could I do that would be free to prospects or existing customers that would be likely to further start a sales process?

Don't offer something for free simply because "many things on the Net are free." Anything you give away needs to be part of a well-thought-out plan that concludes with profitable sales.

Business Plans Are
More Essential than Ever

As the pace of competition has accelerated, product life cycles have, in some instances, shrunk from two years to six months; windows of opportunity open and close with blinding speed; and customers, who are constantly being wooed by the competition, are more demanding than ever. Some influential business thinkers, most notably Regis McKenna in his book *Real Time*, seem to suggest that business plans are no longer relevant and that it's impossible to plan for the future.

In fact, in the era of HyperWars, the exact opposite is true. It is now more important than ever before to have a battle plan during what might, in retrospect, be viewed as "peacetime."

Consider what's happened within corporations in the past fifteen years. In the throes of tremendous upheaval, companies have downsized, replaced guaranteed pensions with self-managed 401(k) plans, and made it very clear that employees can no longer anticipate lifetime employment. In this climate of new uncertainty, business and financial experts deliver in unison one consistent message to workers: "Even though you don't know what you'll face in the future, planning and preparation are required for financial survival." No one is saying, "The world is more uncertain for you now, so plans are no longer relevant." The truth is, plans are more important than ever. In this regard, businesses are no different from individuals.

Successful companies in the hypercompetitive era are making very clear business plans. Winning companies will have developed a vision of where they intend to be going, and this blueprint will allow them the flexibility to respond appropriately when the unexpected occurs. This chapter examines what's different about hypercompetitive planning and how it can be implemented in your company.

WHAT'S DIFFERENT ABOUT HYPERCOMPETITIVE PLANNING

In the midst of HyperWars, business plans require different attributes. The networking revolution created by the Web for business-to-business activities and business-to-consumer products and services has offered new opportunities but has created a very different business environment.

When I began this book, I anticipated finding an extensive list of ingredients that underlie business plans of companies that are successfully competing in this new environment. However, after reflecting on my own experiences, meeting with many industry participants, and researching the traits of thousands of companies of all sizes, I concluded that there are essentially only three central tenets that are a part of a hypercompetitive business plan:

1. A focus on speed in all of its manifestations
2. The integration of the Web into the core of what the company does
3. A focus on how the company adds value for its customers

THE IMPORTANCE
OF PLANNING FOR SPEED

Successful hypercompetitors recognize that in the emerging era time is an essential element in the competitive battleground, and speed is the central weapon. Like all artillery, speed is an asset when a company is able to employ it in building its own business; it is a liability when an "armed" competitor is moving like lightning to undermine that core business. And whether it's being used for or against you, in HyperWars it's an ever-present element that must be factored into your plans.

Earlier we discussed the value of speed. Planning for speed involves an even broader view. Successful hypercompetitors take the widest possible focus on speed and create their businesses accordingly. Specifically, winning companies in the emerging era build their plans with the following in mind:

- The speed with which they need to bring a product to market
- The speed with which their competitors might introduce a competitive product
- The speed required to improve existing products and bring enhancement (or future generations) of that product to market
- The speed with which the industry, because of the Internet, could potentially be transformed

When these factors are fully integrated into a business culture, they lead to a clear way of determining the intensity at which to approach the market:

1. Faster, better, cheaper . . .

It's getting repetitive, but if a product is faster, better, or cheaper on the Web, companies need to exploit it immediately. Today's mail-order businesses, music and software companies, and all types of niche businesses are going to find their happiest homes on the Web. Specialty businesses such as stock photography companies are running, not walking, to get their catalogs online. Because they foresee that online stock photography shopping will be faster, better, and cheaper, they are investing heavily in the future by planning for it.

2. Get your feet wet now so that you'll be prepared to swim hard very soon.

Even if your competitors aren't there yet, start exploring what the Net can do for your business. Publisher John Wiley has established a new service, Wiley InterScience, that will ultimately offer integrated access to the more than 400 Wiley journals published in disciplines ranging from technology and science to medical, business, and legal professions. Currently in beta test mode, Wiley lets "subscribers" access all information currently online for free. Once all the journals are brought online, it will become a fee-based service.

3. Planned evolution is vital.

Businesses that have the capability to rapidly evolve their products will find it easier to stay ahead of the competition, and developing this capability must be a goal in itself. It is, for example, part of the legend of Silicon Valley that when Hewlett-Packard introduces a new printer, it is already well on the way in developing—and deciding when to introduce—the next *two* generations of printers.

4. Acquisition and integration may be more effective than development.

As detailed earlier, a large measure of Cisco's strategy (and related success) has also been built on the notion of speed. Over the past several years, Cisco has acquired over twenty-five companies and added thousands of employees in an effort to dominate the data networking market. The principal factor driving these acquisitions is Cisco's belief that in order to participate in a market early enough, it's better to acquire a proven entity as opposed to taking the time to develop similar capabilities in-house.

And Lucent Technologies, a company that was spun off from AT&T several years ago, has adopted a methodology similar to Cisco's, acquiring the businesses it needs in order to quickly enter specific markets and stay strong. Partially as a result of its strategy, Lucent is now among the nation's most well-regarded technology companies.

5. Time really is money.

For Cisco, there's no time to waste. Before the ink is even dry on a deal, Cisco has sent in its information technology department to establish an aggressive integration program of the company's new technology.

6. Hypercompetitors can't waste time being concerned about cannibalization.

Worry about "cannibalization" (creating one product to replace another) assumes that a company *owns* a market and has time to leave a product in the marketplace until the company is ready to replace it with something new. Smart companies realize that this is an outmoded way of thinking. The new breed of winning companies assume that the competition is right at their heels and any competitive advantage they have is fleeting. As a result, they don't worry about hurting sales of an existing product by bringing out a new one; they are much more concerned with constantly driving to stay ahead of the competition.

Some brick-and-mortar companies have found it more difficult than others to let go of the "old way" and are sometimes less nimble than newcomers. To survive, they will need to find ways of eliminating excess baggage. Most management thinkers question whether this kind of reinvention is possible. I take a more optimistic view—I feel companies that recognize the need for online success can establish subsidiaries, with an independent management, that can have the freedom and resources to win in a hypercompetitive environment.

Companies that can perform tasks most quickly will be the preferred brands of consumers and in the business-to-business market. Therefore, creating a company that moves like electrons through silicon is paramount. This will determine how well positioned the company is to take advantage of new innovations and compete in the evolving business world.

INTEGRATING THE WEB INTO ALL ASPECTS OF THE COMPANY

"Has our company fully integrated the benefits of the Internet into every facet of our business?" is a question every executive must ask when reviewing a business plan.

A sales Web site cannot be viewed as a separate initiative run by one part of a company. Rather, it must be viewed as part of a total company initiative, seamlessly networked to all of the company's activities. Today companies at the leading edge of Web sales have not only moved a major share of their sales to the Web but also have adapted all of the back-end systems, ranging from ordering supplies to assembling orders, in such a way that they take advantage of the internetworked world. In fact, the "Web-centric" focus is one, often overlooked, reason for Dell's ongoing success.

THE IMPORTANCE OF ADDING VALUE

A good business plan should also answer this question: "In the evolving competitive arena, how does our product or service add value?" Companies that focus on this issue and act on their findings are most likely to develop successful plans that build companies to which the customers remain loyal.

Ingram's Micro, a giant computer products distributor, has proactively developed a plan to remain relevant in this changing world and has moved quickly to implement it. The company is leveraging the Net to redefine the way it brings added value to customers in a way that will permit the company, a middleman, to continue to build customers. The company's ongoing transformation clearly reflects a well-thought-out plan.

The giant computer products distributor is aggressively moving beyond electronic transactions to use an extranet to streamline the sharing of data with product suppliers and customers in order to ease the purchasing process. Creating what they are calling the Business Reseller Center, Ingram's system lets customers enter more sophisticated queries, system configurations, and quotations, all without having to speak to a sales agent.

A second aspect of Ingram's well-orchestrated transformation occurred in early 1998, when the company set up a conveyor belt through its warehouse. Now producing computers for Compaq, IBM, Hewlett-Packard, Apple, and Acer, Ingram has become part assembler, part middleman, and part retailer: After the computer is built, Ingram ships the computer directly to the customer with a label on it indicating that it came from the local dealer.

One might think that relatively young companies could still live by their original value-oriented business goals, but in this super-speed world, nothing can be taken for granted. Consider Intuit, whose history we detailed at the beginning of this book. As software became freely available on the Net, Intuit, known mainly for its packaged software programs, concluded that it needed to rethink its customer appeal in order to maintain a revenue base. Now the company has established relationships with multiple suppliers in order to enhance the quality of its packaged software. Its efforts let customers interface with popular Web sites such as Fidelity, American Express, and Checkfree, an online payment system—services that would be of value to Intuit customers. Intuit has also set itself up on the Web to serve as a clearinghouse for the wide range of financial services that might be sought by individuals or small businesses. Its site conducts product filtering followed by price and product-feature comparisons. Clearly, Intuit has recognized the changes that are coming to its industry, and it is developing a business plan designed to let it continue to provide value.

THE BATTLE PLAN

In the emerging competitive environment, there is one clear blue-print for disaster: to acknowledge the ferocity of approaching change but to conclude, "So I'll deal with it when it arrives." By then others will have positioned themselves to take advantage of these changes. Companies that are "late out of the block" will most likely become also-rans.

Moreover, once the tsunami hits a particular industry, radical change happens seemingly overnight. Companies that have already thought through and planned for potential scenarios will have a tremendous advantage.

So, what's my advice? Ignore the pundits who say that technology is changing so rapidly you can't plan for it. Get busy, develop your plan, and start implementing it now. The thunder that precedes the rainstorm is already being heard; the wise will take along their umbrellas as they step out into this new era.

Flexibility Is Key

HyperWars present a "commitment paradox." Successful companies must be fully committed to a course of action, but at the same time, they also need the ability to recognize when it's time to change. Reacting to the unanticipated will be key to survival, and businesses with strong leadership and a clear focus will be those that succeed in this new era.

America Online is a company that, to date, has not hesitated to shift direction, actually specializing in hairpin turns in order to remain successful. Every year it seems the company has taken a new approach to how it will generate profits, shifting from time-based usage to earning revenue by charging advertisers for access to its customer base.

While companies generally need to lead in change rather than simply react, there will inevitably be times when even the best, most farsighted companies are forced to respond to alterations that were not foreseen, and at that time, two elements become crucial: (1) the flexibility to move in a different direction, and (2) the ability to do so quickly.

By their very nature, speed and flexibility go hand in hand. It's hard to imagine that a firm could operate with speed—as I have defined it in this book—without a fair measure of flexibility.

Since no one can accurately predict exactly how the Web will evolve and how it will impact businesses, companies that can "bob and weave" in response to each new market development will create an advantage in the evolving era of online commerce.

The ability to act with great flexibility includes a number of attributes that are, in and of themselves, important for success. It's these characteristics that are the focus of this chapter.

THE NATURE OF FLEXIBILITY

One can't discuss corporate flexibility without bringing up the name Microsoft. While the company has admittedly missed the timing of some business developments, its flexibility has kept it at the cutting edge of events in the business world today.

Surprisingly, Microsoft was late in getting involved in the Internet. In 1994 Netscape had released early versions of its browser while Microsoft was building the Microsoft Network (MSN) on a proprietary software platform rather than as a portal to the Internet. In a now-famous series of events, Bill Gates came to realize that the Internet was the future, and declared that, henceforth, the Internet would be built into everything the company did.

Rarely does a major company make such a dramatic shift in its business practices. But because flexibility ruled, the company—and Bill Gates in particular—was not afraid to look at the world and effectively say: "We missed something, but we're going to go get it."

Next, the company identified the browser as the central product it wanted to provide. Because Microsoft knew it was late to the table, the company opted to license browser technology from Spyglass, the holders of the original rights to the innovation, rather than develop it on its own. The rest is history being played out on a world stage. At this writing, two browsers, Netscape Navigator and Microsoft Explorer, dominate the market.

However, Bill Gates gained more than just browser dominance when he cut his deal with Spyglass—he learned an important lesson about positioning. Since then, Microsoft has begun a policy of investing in multiple Internet technologies and multiple platforms (such as those based on cable television, over-the-air television, streaming video, etc.). By investing across the board, Microsoft is unlikely to ever again be late out of the starting gate. Whatever technology wins, Gates is already a part of it.

Few companies have Microsoft's resources or the ability to manage experimentation on such a large scale. However, the company's focus on staying flexible in the face of an uncertain future is an example to follow.

The best Web sites use the Net to solicit feedback from customers and change accordingly. Barnesandnoble.com developed a more easily navigable site, and added special features (one-click ordering, a stronger search engine, a software superstore, and revamped subject areas) in response to customer input. Sporting-goods retailer Recreational Equipment Inc. (REI) encourages visitors to send suggestions, and based on this feedback, it has overhauled its Web site.

FLEXIBLE PLANNING

In the emerging competitive environment, it may be impossible to predict what the future will look like, but it is possible to have a feel for a variety of outcomes. As a consequence, it is important to always consider how easily you would be able to shift direction if any one of the outcomes you've considered were to occur. How long would it take to shift gears, and how expensive would it be to do it?

I am often asked, "How much should I budget for the development and launch of a Web site?"

While the appropriate percentage of an overall budget varies according to industry and company, here's what I tell those who ask: The sum allotted should follow what I call "Judson's Rule of Thirds": One-third of your annual budget should go into site development, one-third should go into site promotion, and one-third should be retained for changing the site almost immediately after it goes live. My experience is that as soon as you bring up the site, you will learn so much that you will want to build on that knowledge by adding new and different features. This is an example of *planned flexibility*.

INVEST IN MAINTAINING FLEXIBILITY

In our increasingly uncertain world, it may be worth paying more, in some cases, in order to maintain flexibility. Short-term contracts, extra investment in modular components, and a host of other possibilities that are particular to individual businesses can dramatically increase an organization's flexibility.

As detailed in Strategy 3, there are many ways to cut costs and increase efficiency without signing on for long-term relationships that can ultimately act as encumbrances. Because Web changes occur quickly, companies must remain nimble so that strategies can be modified in a flash.

One of the reasons for Lou Gerstner's success in turning IBM around has been another aspect of what I have termed "planned flexibility." Before Gerstner, the company sold its own products exclusively. Now, IBM will mix-and-match both its own products and those of other companies to provide a complete solution for a client. In effect, Gerstner established a more flexible approach, and the company's greater success is its reward for this flexibility.

EXAMINE THE POSSIBILITIES

With some creative thought, most companies can learn a great deal through well-designed and -executed experimentation. These experiments can be based on the notion of "What if the future unfolds in this way? How would we react?"

General Motors has noted the changes that car-buying services have wrought on the industry, and it has been developing methods that will help it in making future choices. GM, for example, conducted an extensive test in four states of an innovative GM Buy-Power site that, for the first time, gave consumers direct access to dealer inventories, third-party competitive comparison charts, and the ability to communicate online with a dealer. The site also did not lock consumers into searching among specific dealers with geographic exclusivity. GM carefully monitored this initiative, and the success of this experiment led the company to announce that it intends to make the GM BuyPower site available in all fifty states in the first quarter of 1999.

Chrysler similarly established a two-state pilot program that has been successful enough to merit a nationwide launch. Chrysler has included its top dealers in the system by setting up a Web site where the customer can get an online quote from a Chrysler dealer of his or her choice. The dealer must respond within forty-eight hours, informing the customer as to whether the car is in stock or must be ordered from the factory. The initial reports from the California/Maryland pilot program are impressive: 29 percent of consumers who received quotes bought the vehicles, nearly twice the statistical response of a walk-in dealership showroom.

BUILD MODULAR

"Build a business from the ground up" used to be excellent business advice; however, the advice of today should be "Build modular."

Flexibility is particularly daunting if a business is built like a house of cards, with every unit depending on the others. Instead, today's products and services should be designed so that they work together as well as separately. This means that companies can typically respond faster to industry changes because they need only shift certain aspects of their product or service instead of re-creating everything. In designing the Web site that sells my newsletter,

Bruce Judson's Grow Your Profits, I specifically compartmentalized features, all of which gain strength by working together, so that individual items could be easily replaced as the market evolved.

Elsewhere I discussed that IBM revived its PC business, in part, by establishing a planned flexible approach to adding components to different models. This flexibility includes a program where the service business can combine different products for different solutions; this is the core of the success of IBM's service business.

ACT PROACTIVELY

If existing companies make a determined effort to act, they may block new virtual competitors from arising. This may mean protecting their businesses through means that are contrary to what are generally viewed as current best business practices, such as accepting lower prices for Web-based products or lowering prices across the board. However, keep in mind that if you don't do it, someone else may come in and undercut you, or, worse, wipe you out.

During the 1997 holiday season, computer manufacturers found that custom-configured PCs were a necessity for strong sales, and CompUSA is one example where a store "reacted proactively" by deciding to compete against the very manufacturers whose machines it sells. Right next to the displays of IBM, Compaq, and Hewlett-Packard machines in CompUSA stores stands a machine labeled CompUSA. Upon entering a kiosk in the store, customers can configure the exact CompUSA model they want (or go home and do it via the Internet). A computer is delivered within days. CompUSA is attempting to offer the best of both worlds by letting its customers see and touch and ask about what they are buying while still offering the customization customers can now find on the Web. CompUSA also made an aggressive push to build its store brand online. No one can predict whether CompUSA will succeed or not, but the company is clearly determined to pursue multiple approaches and change as its surrounding competitive environment shifts.

RECOGNIZING THE NEED
FOR RADICAL SURGERY

In some instances, the changes brought by the growth of the Internet may be so fundamental that to survive you will need to alter your business. The quintessential example of a product in this situation is the paper-based encyclopedia, which has all but disappeared amid the CD-ROM and online onslaught. In essence, radical surgery is sometimes the only solution for survival.

Long-distance carriers face several challenges to their long-term profitability. On one side, established long-distance services are threatened by Internet telephony. As new devices and technology become available, this super-low-cost means of communicating abroad (through the Internet without long-distance charges) will become more and more of a mainstream business and consumer tool. On the other side, existing long-distance providers are threatened by well-financed start-ups that take advantage of Internet technologies to create lower-cost systems. One example is Qwest, which is presently providing long-distance service in multiple areas and has grown, in part, by setting prices that dramatically undercut Sprint's and AT&T's previously established rate structures.

In an interview in *Wired,* Joe Nacchio, the head of Qwest and a former AT&T executive, expressed how hard radical surgery is: "The telcos realize that a revolution is occurring, but what are they going to do? If they say they are going to cut their margins by 50 percent to compete in this new environment, their stock will drop by 30 points."

Nonetheless, one of the first public acts of AT&T's new chairman, Michael Armstrong, was to announce that AT&T would overhaul its global network and respond to this competitive threat. AT&T

announced that it would offer Internet protocol–based telephony to its long-distance customers and to its WorldNet Internet subscribers at rates comparable to those offered by these new competitors. He subsequently began an even greater transformation, making AT&T one of the largest local cable operators in the nation.

Armstrong recognized that he had entered a HyperWar and he took positive steps toward survival. While Nacchio doubted that AT&T would ever cannibalize its existing business through a lower rate structure, Armstrong saw that such "old-line" thinking was inappropriate for the emerging hypercompetitive environment. He saw the world had changed and that AT&T needed to change with it.

THE ULTIMATE IN FLEXIBILITY: FACING CHANNEL CONFLICT

Conflict among distribution channels (when a manufacturer suddenly starts selling direct or a new type of middleman emerges) is one of the most difficult issues businesses are encountering in this new Net-based economy. While there are no simple solutions, some basic guidelines for approaching channel conflict may help:

1. Companies that do not address the issue are doomed.

Lower-cost direct competitors will inevitably chip away and destroy the businesses of firms that work exclusively through traditional retailers. While many industries may continue to need the regular supply chain, selling direct via the Internet lets a company be increasingly customer-focused. Dell branched out from "simple" direct sales via the Internet to home pages for individual customers that track their specific needs; Dell has also added special features for prospects (as discussed in Part I) to permit them to comparison shop, with the thought that the customer will return to Dell. All of these advancements required the first step of using the Web as a one-to-one sales mechanism.

2. Lowest cost for value wins.

Companies that face potential competitors with lower distribution costs because of direct sales must start this transition as well. To remain viable, companies will need to match the lower cost structures (and hence lower pricing) of these direct-sales competitors. As discussed above, the direct-sales model also engenders a one-to-one sales relationship that can allow the company to add value to the purchase in a way it cannot when retailers are intermediaries.

3. Mixed distribution systems may well be a viable solution for the future.

As each participant in the supply chain redefines its role, new models are likely to emerge. Roles that were once black-and-white may now tolerate many shades of gray—depending on the value each member of the chain provides.

As we go to press, Microsoft is planning an interesting mixed-sales approach: At its main site, visitors will be able to pop items into a shopping cart, but before they get to the "checkout counter" to provide payment information, Microsoft will offer them the opportunity to buy direct from one of its resellers, which in turn can offer the products at significant discounts to the full recommended retail price that Microsoft will charge.

Because retail outlet sales still matter, and companies are loath to give up shelf space or point-of-purchase promotions, Microsoft felt it was important not to undercut retailers. However, it is clearly gaining experience in the event it wants to move to a more aggressive direct-sales strategy.

This is such a significant issue that even some packaged-goods companies are quietly investigating direct Web-based sales. Today, these experiments tend to focus on niche products. Nabisco, for example, announced that "for a limited time only" it was offering Knox NutraJoint gelatin drink mix, a nutritional supplement, direct to consumers at a 20 percent discount. Many grocers don't carry this product, and Knox's Web site stated that visitors can "enjoy the . . . convenience of having Knox NutraJoint delivered directly to your door."

THE DEATH—AND REBIRTH—
OF A SALESPERSON

If you're a middleman, ignore the Internet at your peril, particularly if you sell something that doesn't need to be tried on, tested, tasted, or touched. Otherwise, follow these three guiding principles as you consider ways to reinvent yourself:

1. Focus on what will be, rather than fighting the inevitable.

Distributors and retailers can respond to the emerging online environment in a variety of ways: One of these is to attempt to block manufacturers from adopting new sales models, but the relief will be only temporary. Over the long term, it will be impossible.

2. Examine ways in which you can re-create your business to add value.

Five years ago, Marshall Industries CEO Rob Rodin attended a "Total Quality Management" seminar that got him thinking about the future of his company, the fourth-largest electronics distribution firm in the country (Marshall sells components on behalf of about 150 suppliers). Formerly a traditional middleman, Marshall Industries has transformed itself into a hypercompetitive Web-oriented distribution business. The company's revenues have doubled since it began its shift into e-commerce five years ago, and several hundred million of its $1.5 billion in revenues will be derived online.

Marshall has taken everything it used to do physically and converted it to the Net. Before, an engineer designing a new piece of multimedia hardware would have called Marshall, requested technical literature, reviewed it, and ordered a developer's kit—an interaction that could have taken weeks.

Today an engineer designing something new can visit Marshall's Electronic Design Center, find the technical specs of the part, and simulate designs using these chips. Next, the potential purchaser can download sample code, modify the code to suit this particular product, test it on a virtual chip that's "attached" to the site, and analyze how it performs under these circumstances. If the engineer/customer likes it, within minutes Marshall downloads the code, burns it into physical chips, and sends out samples for designing prototypes. Marshall "wins" by using the Internet to add speed and flexibility to its customer offerings.

Like Marshall, other middlemen have evolved their businesses so that their companies continue to be the best places for buyers to go for products. Intraware, Inc., of Orinda, California, is another middleman intent on survival. To continue to provide additional value, this software reseller created SubscribNet. This service notifies customers about fixes and upgrades and gives them a customized World Wide Web site where the upgrades can be downloaded. Without SubscribNet, customers either would be in the dark about bugs and upgrades or would have to track down the information themselves, sometimes after suffering a computer crash. In this way, Intraware has become a middleman that can make a difference. When the vendors start improving their services and notifying customers themselves, Intraware may need to evolve yet again in order to provide a new value-added service.

And electronics distributor Sager Electronics views the Internet as an opportunity rather than a threat—it intends to use the Internet to help it double its business by the year 2000. Working with its extranet, Sager has developed an "eXtranet eXclusive" designed to give qualified customers real-time access to inventory, customer-specific pricing, and order status round-the-clock, seven days a week. The site also features a catalog where users can perform searches, compare product attributes, request samples and quotes, and place orders. Sager is clearly focused on being well positioned to continue its middleman status.

3. Establish your own strong online presence and brand.

Online retailers can often provide buyers value that a manufacturer cannot: Detailed feature and price comparisons, which are perceived as neutral, among different product manufacturers are one clear example. Earlier in the book, I discussed how NECX, the online computer software and hardware retailer, developed this capability as a compelling attribute to bring customers to its Web-based store.

DEVELOP FLEXIBILITY
THROUGHOUT THE COMPANY

There is one final important point to be made about flexibility: Truly flexible organizations recognize that change must be embraced throughout the firm. Any business that wants to accomplish something significant will need flexibility and support from every department in the company.

Corporate goals are not achieved in isolation: Each element of the company must orient itself to support this objective.

For example, when Rodin wanted to realign Marshall Industries, he looked at the sales-compensation structure and concluded that it would conflict with the overall goals he was establishing. As a consequence and despite extensive objections, he eliminated commissions for salespeople. The strategy worked and Marshall's sales grew significantly.

I have often felt that a company can be like a laser or like a roller-skating rink. In a laser, the light rays are all aligned—providing extraordinary power by supporting each other—and the target of a laser can be redirected instantaneously. In a roller-skating rink, many independent entities are simply moving in a circle while they bump into each other. To be flexible, companies must look at themselves and say, "How can we emulate the laser?"

THE BATTLE PLAN

In my teens, I read an endless number of novels that detailed how the Pentagon planned for enemy attacks of all sorts: With response time limited to a few minutes, the military had to have a thorough plan for its response, no matter what the threat.

As I write this book, my mind keeps returning to these scenes. In today's fast-paced environment, companies will succeed only if they have the flexibility to respond to events of all types. Yet as business becomes ever-more demanding, scenario-based planning is likely to be a low priority.

Nonetheless, I suggest that every businessperson engage in one simple exercise at least once a month: Make a list of the ten ways your business could be threatened, or wiped out, by competitors, both known and unknown.

Then, the critical question arises: How fast could you respond to each of these ten challenges, and would it be fast enough?

Never Stop Looking over Your Shoulder

In the emerging competitive environment, success requires that companies remain ever vigilant. As I have discussed throughout, the new reality of business is that your products and services are always at risk: Whether your business is large or small, you've got to prepare for the day when another company tests your vulnerability. That day could come in three to five years, or it could come tomorrow.

Despite their youth, even online retailers are constantly looking over their shoulders and worrying about the efforts under way by brick-and-mortar retailers as they rise to the competitive challenge presented by the Internet. However, online retailers are well aware that significant competition is also approaching from a different direction: Consumers who use bots are led to the company with the best price, and it is entirely possible that eventually—despite distribution channel conflicts—manufacturers will cut out the middleman altogether and sell directly to consumers. In essence, even the online store may need to go through a major transformation in order to redefine its role in the sales process.

In fact, a recognition of the uncertain nature of the current online retail environment may be what spurred Amazon.com to buy Junglee, a search technology company that powers a number of shopping comparison services. Not only will Junglee allow Amazon to move into the business of selling products in new categories, but it will also create a valuable business and technology base for the company. This provides Amazon with a battle plan if Internet retailing redefines itself in a way that minimizes the importance of an online store.

As the Web obliterates traditional supply chains, competition can come from anywhere, so companies must continually assess where they stand and acknowledge the potential threats to their businesses. No matter how new your company or fresh your ideas, there are no guarantees of permanence in this new online world. This requires businesses of all kinds to adopt radical ideas and strategies in order to stay ahead of the competition. Andy Grove, the former CEO of Intel, said it well when he titled his book *Only the Paranoid Survive*. John Chambers, the CEO of Cisco, similarly confesses an ongoing paranoia about competitors and keeping up with the changing needs of product users.

This chapter illustrates how companies can remain on guard. By staying proactive and aware that at any time there may be changes in the distribution channels, companies can take advantage of new opportunities as they arise and prevent competitors from gaining market share.

BEING AT "FIRST ALERT"

A fundamental aspect of profiting in this intensely competitive world is to constantly maintain an awareness to changes of all types. Some not immediately relevant occurrences may result in an important development for your business or industry. You may be fighting a war on one front against identified competitors when, in fact, the greatest threat to your business is about to emerge on another front from a previously unidentified enemy. Consider what's happening to the telephone companies' advertising books, the Yellow Pages.

Although it is not widely recognized, the Yellow Pages generally contribute a very significant portion of a local phone company's profits, and until recently, each local Yellow Pages had limited competition. However, entirely new entities of all shapes and sizes are now competing with the Yellow Pages for local advertising dollars. Online city guides such as Microsoft's Sidewalk, America Online's Digital Cities, and City Search all have the potential to fill the same need as that formerly filled only by the Yellow Pages. And Bell Atlantic's Big Yellow is an example of a phone company developing a product where one's computer mouse can do the "walking." Bell Atlantic is not only moving to protect its own turf but invading other regions as well. Intense HyperWars in this industry lie ahead.

To maintain turf for your business in this new competitive economy, you need to act as if you are always fighting a two-front war by asking yourself three questions:

1. What is my strategy as opposed to my existing competitors' strategies?
2. Is there anything happening in the external world that could lead someone else to enter my business in a new and different way?
3. Could any recent developments make my business obsolete?

PROTECTIVE RULES TO LIVE BY

To help in making the above determinations, I offer the following six guidelines.

1. If a new technology holds promise, assume that entrepreneurs or competitors will improve it to the point where it is sufficiently reliable to be a serious threat.

At first the ability to route long-distance calls and faxes through the Internet at far lower cost than prevailing long-distance rates was downplayed by major long-distance carriers: "The quality will never be up to the standards needed for reliable voice connections" and "It's good for students traveling abroad who want to call home" were the types of comments made about Internet telephony. All the while, scrappy entrepreneurs continued to refine the quality and develop software to make the service easier to use, and now telephone companies are facing a different world: "The Big Three must be shaking in their boots" and "The explosive growth of the Internet poses a life or death challenge to phone companies. . . ." write analysts today about the fact that entrepreneurs and big players alike are getting involved in Internet telephony. The long-distance carriers need to be watching out from all sides, and the technology that will affect them is changing daily, meaning that every day can bring new developments.

Consider, too, what's happening to phone companies in the facsimile market: Today about 40 percent of international calls and 8 percent of domestic calls are faxes. This represents $8 billion for AT&T and its rivals—revenue they aren't going to want to lose. Yet digital data transfer, using secure systems such as UPS's or the Lexis Nexis system, has the potential to cut heavily into the fax market.

In each of these cases—and many more—established companies failed to appreciate the significance of the technological changes taking place or the speed with which they would occur. At one level, this is understandable, since the industry giants were facing small, often unknown entrepreneurs. What the large corporations underestimated was that a major weapon (the Internet) in the hands of a small enemy made that enemy a formidable competitor. It has allowed entrepreneurs to exploit new ideas on a scale, and with a speed, that has never existed before.

2. Dropping prices in technology will continue to cause change for core businesses of all types.

Today Kodak's chief rival in film sales is Fuji. Tomorrow Kodak is likely to be embroiled in a whole new battle as low-cost color printers that can print photographic quality become good enough and cheap enough that they are a logical option for consumers. Kodak needs to be asking, "As prices of digital cameras drop and quality improves, will the consumer film market shrink significantly as consumers take digital pictures, store them in virtual albums on their computers and the Web, and print them out when needed?"

Moreover, another industry titan is entering the fray. Hewlett-Packard is moving aggressively into low-cost high-quality color printers and digital imaging devices. Its goal is to give the consumer the control discussed above and take money that would today otherwise be spent on film from Kodak. It's worth noting that, to date, Hewlett-Packard has not generally been regarded as one of Kodak's principal competitors.

The ultimate resolution of this hyperbattle cannot be predicted. The more important point, however, is that such a skirmish is brewing. Increasingly low costs for technology, combined with the capabilities of the Internet, mean that established companies that previously coexisted peacefully may suddenly be pitched into battle with each other. And, of course, it is virtually impossible to anticipate the other developments that could bring new competitors into this multifront war.

3. Dominant players in one market can successfully be beaten by players from other markets.

Can Amazon sell music as well as it sells books? The company certainly intends to give the virtual music industry a run for the money. As an established trusted retailer with an ability to recommend particular products for particular tastes, it may pose a real threat to the dominance of today's leading online music stores.

Yahoo!, arguably the most well-known brand in cyberspace, will be punching left and right in an attempt to maintain market dominance as a portal for visitors entering the Web. Other search engines, AOL and Microsoft, arrived to fight the good fight, and now established media companies such as NBC and Disney also want to be portals for entry.

The best way to minimize this ever-present threat is either to work to become the low-cost provider or to build loyalty into your customer base through personalized services—and sometimes both.

While the newspaper industry has been smart and quick in finding ways to maintain revenue from the classifieds, the fact that so many different types of companies, from CarPoint (Microsoft) to Realtors.com (affiliated with the National Association of Realtors), are entering the business means that the battle is far from won.

As you consider your own company and its traditional competitors, remember to look outward from that horizon. The Internet may be creating competitors that never could have challenged you in the physical world.

4. To thrive in this intensely competitive world, there is little time to rest on the laurels of a successful product. Companies must constantly be asking, "How fast can we make this product even better?"

To the maximum extent possible, you want to keep moving forward to prevent your competitors from getting a strong foothold in any of your markets. If you can prevent a competitor from establishing a beachhead in an evolving market, you may prevent it from moving inland toward your base.

Remember, too, that competitors are unquestionably watching your every move. If they enter the market with a "me too" strategy, they will likely do so with some type of enhancement, taking their products beyond yours. This then places you, the market leader, as number two in a game of catch-up, a position you never want to play.

Companies may find it's faster, easier, and better to acquire companies with the skills and technology needed rather than attempt to develop it themselves. In an industry publication, Cisco's CEO John Chambers talks of one product the company had in development only to be told by customers, "We're going to make decisions over the next year or two for vendors who have the product available now. If you don't have it, you can't participate." Chambers's response was to go out and buy a company that already had the product in question.

5. New digital age middlemen may appear at any time.

Many businesses rely on the notion that manufacturers or buyers would never go around them because it would alienate their entire distribution chain. As discussed previously, an important component for businesses that operate brick-and-mortar or online stores is that ultimately the lowest-cost distribution method generally wins. Many companies will inevitably find that their traditional supply chains will go through upheaval.

It's safe to say that if you operate in a niche of almost any kind, someone is probably thinking about, or already creating, a new middleman that will have an impact on your business and your industry. It has happened in businesses as varied as financial services, the wedding industry, and polymer manufacturing.

In fact, it's possible that the entire concept of stores on the Web will go through a major transition. Today online emporiums largely resemble their physical counterparts and are the middlemen of cyberspace. However, they, too, must keep looking over their shoulders. In some industries or market segments, they may be replaced by bots that are ready to take the place of online middlemen. Current online enterprises must continually ask themselves if they are adding enough value to maintain their customer base.

Eventually, however, the inevitable will occur. As bots become more capable and consumer-friendly in searching out the lowest prices for products on the Web, it's hard to imagine that they won't wrest a great deal of the business away from online stores, permitting manufacturers to sell direct. At this point, however, the manufacturer faces new challenges, including distribution, fulfillment, and billing. So, it may be that stores evolve to fill the gaps that are left open by the manufacturer. This model already exists, to some extent, where service companies help software makers distribute product directly to consumers.

Because online businesses are just beginning, the process is filled with the unknown. Jeff Bezos, the founder and chief executive of Amazon.com, has described it as the "Kitty Hawk era of electronic commerce." In this age, companies must constantly *earn* their markets and their customers.

With this in mind, businesses will constantly be moving to develop services that continuously add value in the face of new developments. Firms will also increasingly be fighting multifront wars: Their identified competitors (both on- and offline) and potential new competitors, which may even include their suppliers or distributors.

6. Do everything you can to build loyalty into your customer base.

Personalized selling and the "total solution" (as discussed in Strategies 5 and 6), with personalized recommendations, customized products, and added service features, both add value for customers. As explored earlier, companies as diverse as Office Max and Garden Escape have created tools that naturally enhance loyalty. Quicken is hoping to provide one-stop shopping in order to build loyalty.

The winners in these battles will be those who follow the strategies laid out here: Be first, be the low-cost provider, market aggressively, and work to build customer loyalty through personalized features or customized offerings.

THE BATTLE PLAN

Recently I've heard CEOs from a few companies describe themselves as "more paranoid than Andy Grove." In the past, paranoia was generally viewed negatively. Certainly, a corporate board would have thought twice about installing a leader who viewed paranoia as an important and valuable part of his or her approach to business.

I believe that the emergence of this new accolade reflects the reality of business today. Companies are constantly facing threats from known competitors that are at their heels as well as from innovative companies that may transform an industry. In this climate, it is critical for success that the leaders and members of each organization manage to "keep looking over their shoulders" even as they work to move forward.

To put this strategy to work for your business, keep asking yourself questions that start with "What if" about potential competitors. Then follow up by asking: "Am I proactively protecting my turf by continuing to build my business?"

This, in turn, should be followed by: "Am I doing all I can to ensure that if that threat emerges, my business can still thrive?"

Paranoid? Maybe. Smart? Absolutely.

TOWARD THE FUTURE:
A PERSPECTIVE

started this book by reflecting on my experiences and discussing my thinking with industry leaders, many of whom are close friends. When I outlined the ideas for this book, I felt good about them. Now, several months later, I feel even better. Not just because I have finished this chronicle (although that is certainly a source of satisfaction), but because events are already unfolding in line with the ideas and recommendations that are put forward. I have even more reason to believe that the ideas and principles presented will not only continue to shape success in business but will become increasingly important.

My work, both as a business pioneer and a writer, has given me a very particular perspective on how to think about the Internet and the future of business. Now I'd like to leave you with five primary thoughts to keep in mind as you observe what is happening to the business world today:

1. The Internet is so powerful because of a rare combination of the following factors:

- a standard for which anyone can develop products
- a standard that is not owned or controlled by anyone

Many people decry this Wild West–like environment, pointing out that a lot of awful material is developed. But I believe this is a main source of its strength. Since there is no gatekeeper, hundreds of thousands of people are dedicated to entrepreneurial Internet-related activities. In fact, this may be the single greatest rush of entrepreneurial effort in history.

All of this activity means that a lot of useless or awful services and material are created, since people can effectively put anything on the Web they want. At the same time, companies—both large and small—are constantly pushing the envelope of what can be.

The end result: Internet users allow sites that are not useful to become ghost towns, since no one visits them, while valuable innovations occur at an almost inconceivable rate.

2. At one time, I believed that the pace of innovation would slow down. I now believe that the opposite is true: It's likely to become even faster.

My initial reasoning was that as each improvement required larger teams and more effort, we would see a leveling off of innovation. In fact, as the Internet has matured, innovative businesses of the type described in this book can be created with increasing ease. They no longer require the same amount of custom software that would have previously been needed. Instead, they can make use of robust "off the shelf" solutions. This suggests to me that the innovative pace of new business activity on the Internet is more than likely to accelerate.

I have also not seen any slowdown in technical developments. As the Internet has matured, it has also become a focus of the efforts of our nation's leading technology companies, meaning that an almost limitless amount of resources is now available to solve technical issues.

For businesses, this means that the hypercompetitive environment, which spawns HyperWars, is here to stay.

3. Pricing will definitely be affected.

There is a growing belief within the business community that the Internet will fundamentally change the way goods and services are priced. Several factors are typically cited as evidence of this change:

- Comparison sites and bots that find the "best" prices for items on the Web are becoming more prevalent and technologically better, placing intense pricing pressure on producers. As producers grapple with this phenomenon, a number of industry experts believe they are also likely to experiment with new pricing models where they can charge more under different circumstances.

- The popularity of online auctions is growing. Today these auctions deal primarily with surplus goods, but will this continue unchanged? Could auctions, in some form, become an accepted mechanism for selling newly released products to the public?

- Services such as Priceline.com (where the consumer sets the price he or she is willing to pay for airline tickets) are transferring greater purchasing power to the consumer. These services effectively let the consumer say, "I am willing to pay this amount for this flight or this item. Does anyone want to sell it to me at that price?"

- The Internet's ability to gather and disseminate information at a previously unheard-of pace enables companies to change prices rapidly in a way not possible before. Like the airlines that alter prices based on seating inventory, other companies will be able to adjust their prices, too.

Companies that survive in a HyperWars environment will be highly flexible and able to adjust prices far more rapidly than in the past. They will also build their businesses around this notion. In the past, companies introduced hard goods products, expected retailers to sell them at full price for some time, and then fully anticipated that stores would start discounting the items. At some point, the company might discontinue the product and attempt a variety of ways to sell its remaining excess inventory. In the emerging environment, winning companies will be attuned to what is happening in their markets and adjust their prices accordingly. This may mean that suggested retail prices fluctuate frequently, and that the prices of items bought direct from the manufacturer move up and down in response to a variety of factors: product superiority, competitors' prices, overall demand, and short-term product availability.

Does this mean an end to fixed pricing? It likely won't be eliminated altogether; however, the factors listed above combined with the hypercompetitive climate of today make it even less likely that companies will be able to set one price and stay with it through the life of a product.

4. If you can imagine it, it will probably happen.

The Internet today is filled with innovations that were once deemed impractical or "years away from development." These include phone calls with good sound quality based on Internet technology, software that automatically updates itself, bots that accomplish useful tasks, highly valuable comparison-shopping services, voice recognition activities, and all manner of businesses focused on vertical (industry-specific) niches. As you know from reading *HyperWars*, many of these changes that were once almost unimaginable are now well on the way.

For the businessperson, this means you can't ever dismiss the possibility that new competitors or threats will emerge based on technology that "seems" years away or as if it would be impractical. There are an army of people and a flood of resources at work turning the impractical into the routine.

5. If you can imagine it, it probably already exists somewhere.

Time and time again, I have found myself saying, "There should be something that . . ." only to discover a few weeks later that, in fact, what I was envisioning had already been developed, often in secret, and was then unveiled. In some cases, these new developments helped my business, in others, they represented a new threat.

Again, this reflects the huge resources now aligned in this arena. For a businessperson, it is a frightening idea: Someone could come along tomorrow with something that will severely cripple your business. Unfortunately, this is a fact of life in the new era of HyperWars. More importantly, the strategies included in the book are designed with this perspective in mind. Keep this book handy.

AS YOU CLOSE THIS BOOK . . .

When I started discussing this book project, I was constantly asked one question: "How can you possibly write a meaningful book about a world that is changing so quickly? Won't it be instantly out-of-date?" My answer was twofold:

First, I have included different types of information here. The most important parts of the book are the basic business principles I know will hold true in the evolving business climate. To bring these principles to life, I have included a wide variety of examples that cut across all types and sizes of businesses. The goal in providing these examples is to show, in a concrete way, how the governing principles can be put to work. Over time the examples will inevitably become less current, but the principles will continue to apply. The value of the advice shouldn't diminish for a long time to come.

Second, I recognize that many readers may be interested in how some of the events I describe here are continuing to unfold. If you are looking for publications that will keep you up-to-date on developments in this area, I recommend *Internet Week* and *Internet World*. Both are trade publications, but subscriptions are available to individuals. In addition, I have created a Web site and an e-mail newsletter that will, in a sense, continue this book by taking advantage of the Web and the Internet. At the site, **www.hyperwars.com,** I will provide continuing information and ideas on how to steer the course of your business. At the same time, I have also created a capability at the site for readers to send me questions. I will answer as many questions as possible and post the questions and answers weekly so that visitors to the site will have access to a growing electronic reference library. I encourage you to visit the site and to sign up for the free e-mail newsletter.

PART IV

THE HYPERWARS AUDIT: IS YOUR BUSINESS READY FOR BATTLE?

The purpose of this audit is to bring to life, for your company, the ideas discussed in this book. The audit is designed so that in a quick and handy fashion you can see for yourself how prepared your business is to survive and prosper in the emerging hypercompetitive environment.

SECTION I: STATE OF MIND

Competing in a HyperWars environment requires a strong commitment to action. This set of questions is designed to probe how deeply people in your business have adopted the attitudes that will be necessary to succeed on the emerging competitive battleground.

1. What is the mind-set in your company today regarding the Internet and the speed of change?

- By and large, do people within your company view the Web as something that will dramatically change their business, or is it generally regarded as "not a big deal"?

- Where the Internet is regarded as important, do the people in your company believe that they have to act fast to win or lose with regard to the opportunities it presents? Or is the Internet seen as a factor that can gradually be addressed over time?
- Are Web-based businesses considered possible threats to your core business? If not, can you imagine a scenario where they might be?
- Is there a belief within your company that, under certain circumstances, the Internet could transform your basic business at a dizzying rate? If not, what protects your business from the possibility of rapid change?

SECTION II: ASSESSING YOUR READINESS TO COMPETE

A proactive approach and an ability to move quickly and flexibly will be central attributes of winning companies in the era of HyperWars. This section helps you to determine whether your company has sufficiently developed these capabilities.

2. Is speed (to develop new products, to deliver made-to-order goods, to deliver shipments ordered) a priority in and of itself within your company?

- Do you know of companies in your industry that can do things faster than you? If so, why do they have this edge?
- Would objective third parties view your business as nimble?

3. Have you established, and taken full advantage of, a company-wide intranet?

- Is it used to further aid in the overall speed of everything that happens within the company?
- Has it been deployed to cut costs wherever possible?
- Is it used to promote information-sharing that leads to better products or service from the company?

4. Have you established an extranet that encompasses your key customers and suppliers?

- Can you imagine any ways in which the extranet could further build relationships with customers and suppliers, speed development times, or cut costs? If so, have the necessary changes not been implemented for valid reasons?

5. Have you investigated new procurement methods, such as digital age middlemen and preapproved online supply catalogs, that can lower your costs?

SECTION III: BUSINESS VULNERABILITY

Some businesses are inherently more vulnerable to Internet-based onslaughts than others. This section is designed to help you think through the degree of your business's vulnerability to rapid Internet-based change.

6. Can you imagine how any aspect of your product could be delivered better, faster, or cheaper online?

- Can you imagine how online activities might add value to your customers' product experience?
- If you can imagine these possibilities, have they been implemented by your company?
- Have competitors implemented any of these ideas? If so, are you concerned about what these competitors are doing? If not, why not?

7. Can you imagine the development of a service or product that could entirely destroy your business?

- If so, what would the characteristics of this service or product include?
- If any variation of such a service or product were launched,

do you know how you would want your business to react? Do you believe your business could react fast enough in the way you would like it to? If not, should you invest now in developing this type of response?

- Are there actions you can take today to block the successful implementation of a service or product that could destroy your business? Does it make sense to invest now in such efforts?

8. Are you a middleman of any kind (brick and mortar or online, dealer, retailer, agent, reseller, or distributor)?

- If so, are there special ways in which you add value for your customers?
- Do you believe this value is sufficient to prevent customers from buying directly from your suppliers, or to prevent suppliers from selling directly because they know they can't compete with the value you add?

9. Do you sell your product to middlemen (as defined above) or directly to the end user?

- If you do not sell directly to the end user, could you via the Web? Have you started serious experiments to build your business through this distribution channel?

10. Have you experimented with developing new, Web-based products? Are these products targeted at your traditional customer base or at a new base of users? Is it possible to expand your customer base?

SECTION IV: MAKING IT HAPPEN

This section, in a sense, returns to the start of the audit. My research indicates that companies that succeed in adapting their businesses to meet online competition have a game plan and people in place with designated responsibilities.

11. Who in your organization is accountable for the answers to the above questions?

- Do the individuals you believe to be accountable unambiguously recognize that this is an important part of their jobs?

Web Site Addresses of Selected Companies and Services

Adaptec, Inc.	www.adaptec.com
AIG (at Auto-By-Tel)	www.autobytel.com (click on "insure")
AltaVista	www.altavista.com
Amazon.com, Inc.	www.amazon.com
American Airlines, Inc.	www.aa.com
American Finance and Investment Co.	www.loanshop.com
American Girl (Pleasant Company)	www.americangirl.com
American Greetings, Inc.	www.americangreetings.com
America Online, Inc.	www.aol.com
Ameritrade, Inc.	www.ameritrade.com
AMP, Inc.	www.amp.com
Applied Materials, Inc.	www.appliedmaterials.com
Aptex Software, Inc.	www.aptex.com
Atlas Van Lines, Inc.	www.atlasvanlines.com
AT&T Corp.	www.att.com
Auto-By-Tel	www.autobytel.com
barnesandnoble.com, Inc.	www.barnesandnoble.com
Bay Networks, Inc.	www.baynetworks.com
Better Business Bureau OnLine	www.bbbonline.com
Big Yellow (from Bell Atlantic)	www.bigyellow.com
Book-of-the-Month Club	www.bomc.com
BotSpot, Inc.	www.botspot.com
Brightware, Inc.	www.brightware.com
Broderbund Software, Inc.	www.broderbund.com

Note: Readers should be aware that Web site addresses change frequently. The addresses above are accurate as of the time of this writing.

Bruce Judson's Grow Your Profits	www.growyourprofits.com
Burn Rate	www.burnrate.com
Business Week	www.businessweek.com
BUYCOMP, LLC	www.buycomp.com
Calla Bay (swimsuits)	www.callabay.com
Career Path	www.careerpath.com
CarPoint	carpoint.msn.com
Catskill Casket Co., Ltd.	simpworld.com/casket/main/menu.htm
CDNow, Inc.	www.cdnow.com
Charles Schwab & Co., Inc.	www.schwab.com
Chemdex Corp.	www.chemdex.com
Chip Shot Golf Corp.	www.chipshot.com
Chrysler Corp.	www2.chryslercorp.com/default.html
CitySearch, Inc.	www.citysearch.com
Cisco Systems, Inc.	www.cisco.com
Claritin (rebates)	www.claritin.com/rebate/index.htm
Classified Ventures	www.classifiedventures.com
Clinique	www.clinique.com/main.html
Club Med	www.clubmed.com
CNET, Inc.	www.cnet.com
CNET Shopper.com	www.shopper.com
Compaq DirectPlus	www.directplus.compaq.com/productlines.cfm
Compaq Online Solutions	www.clubweb.com/mf_services_onlinesvcs.asp
CompUSA	www.compusa.com
Craftsman (from Sears, Roebuck & Co.)	www.craftsman.com
Daimler Benz	www.daimlerbenz.com/index_e.html
Dell	www.dell.com
Digital City, Inc.	home.digitalcity.com
Direct Stock Market, Inc.	www.directstockmarket.com
Echomail	www.echomail.com
Egghead.com	www.egghead.com
Electronic Newsstand	www.electronicnewsstand.com
Eli Lilly and Company	www.elililly.com
Entrepreneur	www.entrepreneurmag.com
ErgoTech's E-mail Room	www.e-mailroom.com
Ernie (from Ernst & Young LLP)	ernie.ey.com
E-Stamp Corp.	www.estamp.com
eToys, Inc.	www.etoys.com
E*TRADE Securities, Inc.	www.etrade.com
Excite Product Finder	jango.excite.com/xsh/index.dcg?
Fidelity Investments	www.fidelity.com
Ford Motor Company	www2.ford.com
Forrester Research, Inc.	www.forrester.com
Fortune	www.fortune.com
FreeMarkets OnLine, Inc.	www.freemarkets.com
Fruit of the Loom, Inc.	www.fruit.com
Gap Online Store	www.gap.com
Garden Escape, Inc.	www.garden.com
Gateway 2000, Inc.	www.gateway.com
GE Capital Services	www.gecapital.com
GE TradeWeb	www.getradeweb.com

GM BuyPower	www.gmbuypower.com
Grenley-Stewart Resources, Inc.	www.oilprice.com/index.ns.html
Greyhound Lines, Inc.	www.greyhound.com
Gund, Inc.	www.gund.com
Harley-Davidson	www.harley-davidson.com/home.asp
Harrison, Dan (Virtual Pool and Spa Store)	www.paramountpools.com
HP (Hewlett Packard) United States	www.hp.com/usa/buy/
Holland & Hart LLP	www.hollandhart.com
Holt Educational Outlet	www.holtoutlet.com
Hotbot Shopping Directory	shop.hotbot.com
HyperWars	www.hyperwars.com
IBM Corporation	www.ibm.com
Infoseek Corporation	www.infoseek.com
Ingram Micro Inc.	www.ingrammicro.com
InsureRate, Inc.	www.insurerate.com
Intel Corporation	www.intel.com
Intraware's Subscribnet	www.intraware.com/subscription/subscribnet. html
Intuit Inc.	www.intuit.com
iPrint, Inc.	www.iprint.com
IQVC	www.iqvc.com
J. C. Penney Company, Inc.	www.jcpenney.com
jcrew.com	www.jcrew.com
JEM Computers, Inc.	www.jembasement.com
John Hancock's Marketplace	www.johnhancock.com/marketplace
Kauai Exotix	www.kexotix.com
Killen & Associates	www.killen.com
Knoll Pharmaceutical Co.	www.basf.com/businesses/consumer/knoll/ index.html
Knox NutraJoint	www.knox.com/nj/order.html
Kodak (Eastman Kodak Company)	www.kodak.com
Kodak PhotoNet online	kodak.photonet.com
Konica	www.konica.com
Lands' End, Inc.	www.landsend.com
LendingTree	www.lendingtree.com
L.L. Bean, Inc.	www.llbean.com
Lycos	www.lycos.com
Marine Power (extranet)	www.marinepower.com
Marshall Industries	www.marshall.com
McAffee Associates (Network Associates)	www.mcafee.com
McGraw Hill's Primis Custom Publishing	www.mhhe.com/primis
Merrill Lynch & Co., Inc.	www.ml.com
MetLife Online	www.metlife.com
Microsoft HomeAdvisor	homeadvisor.msn.com/nsm/default.asp
Microsoft Shop	www.microsoft.com/isapi/referral/default.asp
Midwestock	www.midwestock.com
Money.com	www.money.com
Music Boulevard	www.musicblvd.com

National Semiconductor Corp.	www.national.com
NECX	www.necx.com
NetMarket	www.netmarket.com
Netscape	www.netscape.com
The New York Times	www.nytimes.com
Nike, Inc.	www.nike.com
No Brainer Blinds and Shades	www.nobrainerblinds.com
Northern Light	www.northernlight.com
Online Weight Loss Clinic	www.practicalprogram.com
Open Market	www.openmarket.com
PricewaterhouseCoopers Tax News	
Network	www.pwcglobal.com/gx/eng/ins-sol/ spec-int/know-dir/tax-news.html
Pall Corporation	www.pall.com
Pitney Bowes Personal Post Office	ww1.pb.com/sohoshop/ppo_home.asp
Polk Audio	www.polkaudio.com
Polymerland (from GE)	www.polymerland.com
Preview Travel's Farefinder	farefinder.previewtravel.com
Preview Travel's Business Travel Center	business.previewtravel.com
priceline.com, Inc.	www.priceline.com
The Prudential Insurance Company of America	www.prudential.com
Quicken.com	www.quicken.com
Qwest Communications International Inc.	www.qwest.com
Realty.com	www.realty.com
Recreational Equipment Inc. (REI)	www.rei.com
Reebok International, Ltd.	www.reebok.com
ResponseNow	www.responsenow.com
Ross Controls, Inc.	www.rosscontrols.com
RSNAlink	www.rsna.org
Sager Electronics	www.sager.com
Security First Network Bank	www.sfnb.com
Security National Mortgage Corp. (RateTracker)	www.lowestrate.com/tracker.shtml
Sidewalk (from Microsoft)	www.sidewalk.com
Standard & Poor's Personal Wealth	www.personalwealth.com
Suretrade.com	www.suretrade.com
Thompson Investor Network (TIN)	www.thomsoninvest.net
Ticketmaster	www.ticketmaster.com
Time Warner's Pathfinder	pathfinder.com
The Tide ClothesLine	www.tide.com
Toshiba	www.toshiba.com
United States Postal Service (USPS)	www.usps.com
Visa International	www.visa.com
Warner Lambert's alergy-cold.com	www.allergy-cold.com/productmatch.html
The Washington Post	www.washingtonpost.com
WebTV Networks, Inc.	www.webtv.com
Wickes Lumber Company	www.wickes.com
Wiley InterScience	www.interscience.wiley.com
Wireless Dimension	www.wirelessdimension.com
Yahoo!	www.yahoo.com

Index